MAN IN THE COLD

Publication Number 986

AMERICAN LECTURE SERIES®

A Monograph in

The BANNERSTONE DIVISION *of*
AMERICAN LECTURES IN ENVIRONMENTAL STUDIES

Edited by

CHARLES G. WILBER
Professor of Zoology
Colorado State University
Fort Collins, Colorado

Man in the Cold

By

JACQUES LeBLANC
Professor, Department of Physiology
School of Medicine
Laval University
Quebec, Canada

With a Foreword by

Charles G. Wilber
Professor of Zoology
Colorado State University
Fort Collins, Colorado

CHARLES C THOMAS · PUBLISHER
Springfield · Illinois · U.S.A.

Published and Distributed Throughout the World by
CHARLES C THOMAS • PUBLISHER
BANNERSTONE HOUSE
301-327 East Lawrence Avenue, Springfield, Illinois, U.S.A.

© *1975, by* CHARLES C THOMAS • PUBLISHER
ISBN 0-398-03429-X
Library of Congress Catalog Card Number: 75 4740

With THOMAS BOOKS *careful attention is given to all details of manufacturing and design. It is the Publisher's desire to present books that are satisfactory as to their physical qualities and artistic possibilities and appropriate for their particular use. THOMAS BOOKS will be true to those laws of quality that assure a good name and good will.*

Printed in the United States of America
N-1

Library of Congress Cataloging in Publication Data

LeBlanc, Jacques, 1921—
    Man in the cold.

    (American lecture series; publication no. 986)
    Bibliography: p.
    Includes index.
    1. Cold—Physiological effect. 2. Cold adaptation.
I. Title. [DNLM: 1. Acclimatization. 2. Cold. 3. Cold climate. QT160
    L445m]
QP82.2.C6L4   612'.01446   75-4740
ISBN 0-398-03429-X

*This book is dedicated to
all my experimental subjects*

FOREWORD

IN THIS VOLUME Professor Jacques LeBlanc brings to bear on the subject of man in the cold a wealth of firsthand experience as a scientific investigator in the field. As a teacher in a medical school he has developed a keen integrative skill which is reflected in this book.

The ever-increasing interest in man's interrelationship with the environment makes this monograph timely. It is important to emphasize in this connection that to observe man in the cold, it is essential to study *man* in the cold, for "recent studies indicate that not all of the adaptive responses, such as biochemical changes, seen in temperature-acclimated rodents can be extrapolated to other species, such as primates" (Chaffee and Roberts, 1971).

Much research on the physiology of cold exposure has involved studies on the ubiquitous laboratory rat. Extrapolations from such studies to man must be done with caution. Dr. LeBlanc has based his work for the most part on human observations.

The present volume should serve as a summary statement of our present understanding of human functional responses to cold exposure. It should help the reader tie into the overall pattern of human responses to cold such observations as local changes in fat composition, changes in amounts and distribution of isoenzymes, modifications of gluconeogenesis—all observed in man exposed to cold.

It is clear that an exciting area for continued research is the clarification of the neurochemical mechanisms involved in the hypothalamic thermostat.

CHARLES G. WILBER

REFERENCE

Chaffee, R.R.J. and Roberts, J.C.: 1971. Temperature acclimation in birds and mammals. *Ann Rev Physiol, 33*:155-202.

CONTENTS

                                                                          Page
Foreword vii

Chapter

1. GENERAL RESPONSES TO COLD 3
   Thermal balance 3
   Temperature regulation 10
   Individual factors in cold tolerance 15
2. METABOLIC EFFECTS OF COLD 25
   Source of energy for thermogenesis 25
   The role of lipids 27
   The importance of brown adipose tissue 32
   Caloric requirements in the cold 35
   Common cold and vitamin C 40
   Cold exposure and vitamin C 44
3. ENDOCRINES 49
   The thyroid gland 49
   The adrenal cortex 58
   The pituitary gland 60
4. CARDIOVASCULAR RESPONSES 63
   Cutaneous circulation in the cold 64
   Systemic cardiovascular response in the cold 64
   Effects of cooling face and hands on systolic time
   intervals (STI) 68
   Individual variations in responses to face or hand cooling. 76
   Cold morbidity 81
   Responses to heat in a cold environment 84
5. ADAPTATION 90
   I—Adaptation to cold in laboratory animals 90
       Adaptation by continuous exposure to moderate cold. 90

Chapter   Page
          Adaptation by intermittant exposure to severe cold...   90
    II—Adaptation to cold in man 116
          Adaptation by continuous exposure to moderate cold. 116
          Adaptation by intermittant exposure to severe cold.. 126
6. INTERSTRESS ADAPTATION 146
    Cold and altitude 147
    Cold and exercise 153
7. HYPOTHERMIA, FROSTBITE AND TISSUE PRESERVATION 161
    Hypothermia 161
    Frostbite 166
    Preservation of Tissues by Hypothermia 169
8. ASSESSMENT OF THE ENVIRONMENT BY RESPONSES
    OF THE FACE 172
    Temperature of the face 172
    Heart rate 175
    Thermal comfort 177
    Windchill value and physiological parameters 181

Index 193

MAN IN THE COLD

GENERAL RESONSES TO COLD

THERMAL BALANCE

TOLERANCE TO COLD is very complex and shows great variations within the different animal species. While Arctic mammals possess a very marked resistance to cold (Fig. 1-1), other species including rats, mice and men have very low capacity for withstanding low temperatures (194). These differences can be explained primarily by physical factors.

The first and most important factor is that of heat loss. The amount of heat which is lost in the cold is proportional to the surface area per unit of body mass. For that reason, independent of other factors such as insulation, smaller animals are much less resistant to cold than larger ones. But this factor is not the only one to consider. Man is about 200 times larger than the rat and yet their respective tolerances to cold may not be different as Figure 1-2 would seem to indicate (96). In a cold environment, rats do not survive temperatures below −15 to −20°C. The lethal temperature for nude man in the cold is not known. However, assuming that the total capacity for heat production (expressed per unit of body mass) is the same in the rat and in man, no proof to the contrary being known, then by referring to Figure 1-2, one could conclude that the lethal temperature for both species is comparable. Considering the amount of heat that man produces compared to rat, it can be said that man has a low resistance to cold; indeed he may be truly compared to a tropical animal. The relatively high degree of resistance to cold observed in the rat is due to its insulation as is well illustrated in Figure I-3 (194). A direct correlation is found between the thickness of the fur and its insulative value. The value of

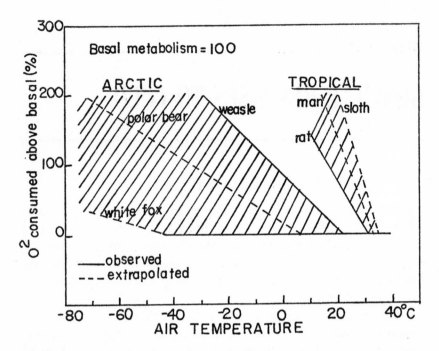

Figure 1-1. Increase in oxygen consumption for different species exposed at various air temperatures. On the basis of the responses the species are classified into a tropical or arctic group; with this criterium man can be considered a tropical animal (194).

insulation in resistance of man to cold has been extensively studied during the second world war. Burton and Edholm have reviewed the literature on this matter and their book has remained a classic on the subject of man in a cold environment (25).

These authors have suggested the *clo* unit for measuring insulation. One clo of thermal insulation will maintain a resting-sitting man, whose metabolism is 50 Kcal/sq.m/hr, indefinitely comfortable in an environment of 21°C (70°F), relative humidity less than 50 percent and air movement 200 ft/min. In these basal conditions one clo is equivalent to a business suit or one quarter inch clothing. A formula has been derived which gives the number of clos required for any given temperature.

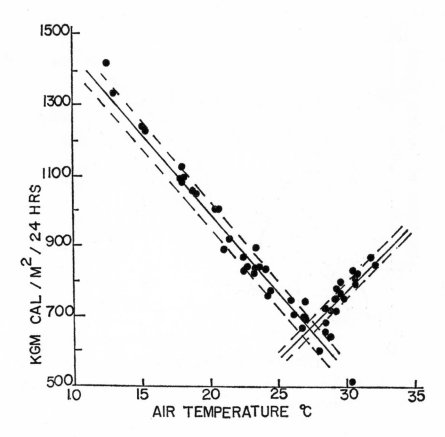

Figure 1-2. Metabolic response to various ambiant temperatures in the rat (96).

$$I = \frac{3.1 \ (T_s - T_a)}{3/4 \ M}$$

I = insulation in clo units.

T_s = average temperature of the skin (33°C or 91.4°F is conveniently taken as comfortable skin temperature).

T_a = temperature of the air.

M = metabolism; 50 kcal/m²/hr is resting metabolism.

Of this heat production 75 percent is lost through clothing and the

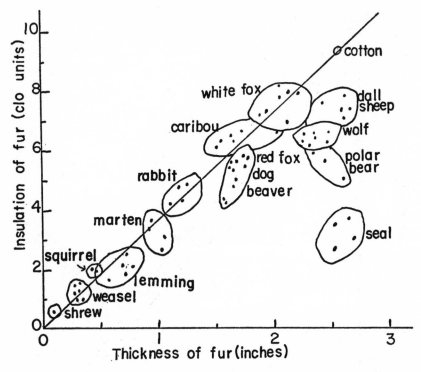

Figure 1-3. Insulation of fur of animals, measured on a standard apparatus (194).

rest by evaporative loss. Using this formula we find that the clothing requirement for a subject sitting in a comfortable room at 70°F is one clo.

$$1 \text{ clo} = \frac{3.1 \ (93 - 71)}{3/4 \ (50)}$$

At 20°F with resting metabolism, the clothing requirements would be 5.8 clo.

$$5.8 \text{ clo} = \frac{3.1 \ (93 - 20)}{3/4 \ (50)}.$$

Figure 1-4 gives a series of curves which permits a rapid estimation of clo requirements at various environmental temperatures and metabolic activities. Another factor which should be consid-

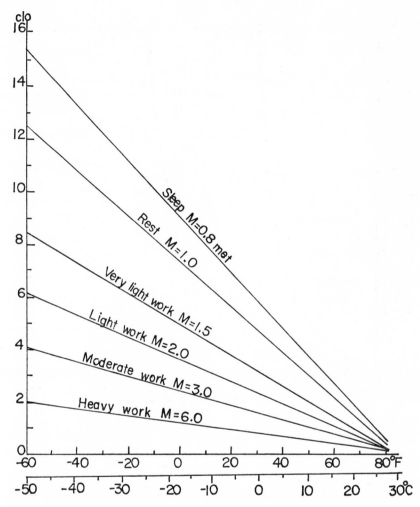

Figure 1-4. Total insulation of clothing plus air needed for different metabolic rates. The formula for the lines is:
$$I = 0.082 \, (91.4 - T°F)/M$$

ered is the speed of the wind. The heat loss to the environment is greatly influenced by wind speed and this effect depends also on metabolic activity. Figure 1-5 gives the temperature decrement which should be subtracted from the actual thermometer reading

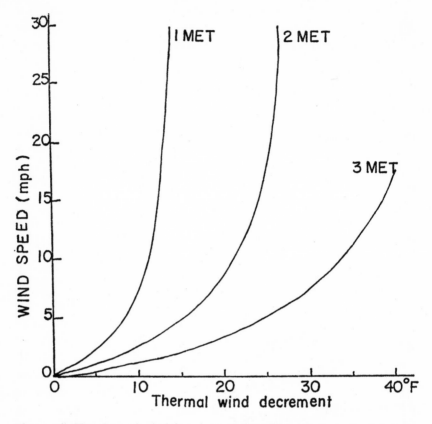

Figure 1-5. The thermal wind-decrement to be subtracted from the thermometer reading to give the equivalent still-air temperature. Note that it depends on the metabolic activity of the man (25).

to give still-air temperatures corrected for wind velocity. These findings prove quite useful and adequate means of measuring clothing requirements in the cold.

— Throughout the ages man has protected himself against the environment by building proper houses and wearing sufficient clothing, and because of that, marked mechanisms for fighting cold did not develop and the temperature tolerance of man is fairly limited. The Eskimos living in the Arctic illustrate very well this situation. Their clothing is composed of two layers of caribou hides to give a thickness of 2 to 2.5 inches or 8 to 10 clos. Even when the physical activity is limited to very light work, the Eskimos wearing this type of clothing can remain comfortable at temperatures as low as —40 to —50°F. Consequently the best way to resist cold is to wear adequate clothing.

The type of clothing should vary with the temperature, the wind, the amount of sunshine and the level of activity. A windproof jacket may be useful in open areas, but an essential quality of a good garment is that it be able to "perspire." The accumulation of sweat in the layers of clothing can be very detrimental for a person who stops exercise or is exposed suddenly to colder temperatures. For the same reason a garment which cannot be ventilated has disadvantages. Accordingly a multilayer clothing, which can be removed or added on, and which has the possibility of being opened or closed easily at the neck to eliminate excess heat or conserve it, is the ideal type of clothing to wear in the cold.

This discussion relates only to overall resistance to cold; that is to means by which heat balance can be maintained under different environmental conditions. Another important problem is that of protection of the face and of the extremities in the cold. The hands and feet are much more difficult to protect adequately. Similarly the face is constantly exposed to the environment; face-masks are never satisfactory because of vision obstruction and icing due to vapors of expired air. When proper clothing is worn, the limiting factors in the cold are then the face and the extremities.

An index for assessing the cooling effect of the environment on exposed skin has been devised by Siple (199); it is called the "windchill" index. Its scale, which goes from zero to 2500, expresses the rate of heat loss in kcal/sq. m/hour of cans of water. Although the scientific basis for applying these observations to humans exposed to cold can be discussed and is subjected to limitation, this index has had an immense popularity and certainly corresponds grossly to common observations. Figure 1-6 shows the different combinations of temperatures and winds which correspond to the conditions ranging from comfortable to intolerably cold. This index finds its usefulness as a warning for the danger of frostbite on exposed skin. However, the windchill does not take into

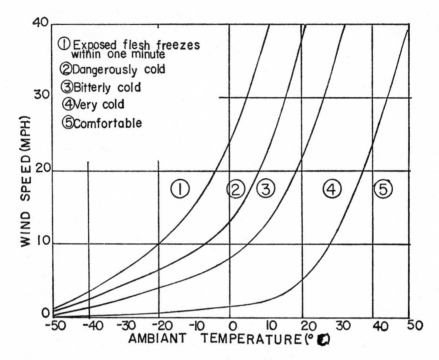

Figure 1-6. Windchill Index for various air temperatures and wind speeds (199).

account the part of the body which is exposed to cold, the level of activity which will affect the heat production, and the amount of clothing which is worn.

TEMPERATURE REGULATION

It has been known for quite some time that spinal cord transection abolishes shivering in the region below the lesion (20). The skin contains a network of fine nerve endings which are specifically sensitive to minute rapid changes in temperature (9, 24). Figure 1-7 shows the impulse frequency of these cold receptors with variations of skin temperatures (92). Rapid cooling increases the frequency of discharge, and warming of the skin completely blocks the increased frequency of impulses normally produced by cooling.

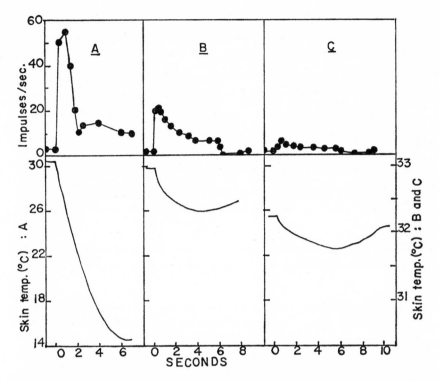

Figure 1-7. Impulse frequency of a single cold fiber in the saphenous nerve of the cat when cooling and warming the skin (fiber no. 3). The conduction velocity of the fiber was 1.5 m/sec. The left hand-temperature scale refers to (A), and the right-hand scale to (B) and (C) (92).

These thermal cutaneous receptors are linked to a central temperature regulating center located in the preoptic region of the hypothalamus. In 1954 Hemingway et al. stimulated different regions of the hypothalamus and midbrain to determine the sites of suppression of shivering. Figure 1-8 shows that stimulation of the hypothalamus, specially in the medial region, completely suppresses shivering measured in the skin of the lateral thigh and in the intercostal muscle (91).

Benzinger et al. (10) have also shown that for a given skin temperature, heat production will be inversely proportional to the

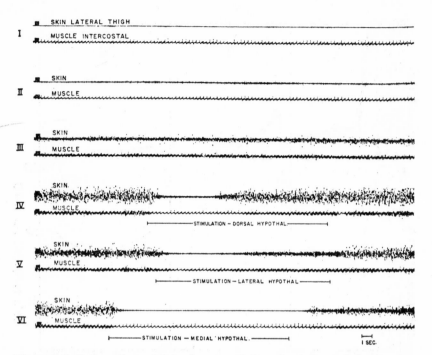

Figure 1-8. Suppression of shivering caused by stimulation of three points in the anterior hypothalamus. The records are monopolar electromyograms with the upper record of each pair being taken from a surface electrode on the thigh and the lower electrode from a needle in the inter-costal muscle. Stimulation of the dorsal hypothalamus records no. IV, caused a brief period of suppression with shivering again starting while the stimulus was on (escape phenomenon). In the lowest record, no. VI stimulation caused suppression during the stimulation and the suppression continued beyond the time of stimuation (after discharge phenomenon). In the upper four records are electromyograms for 3 degrees of shivering. I. no shivering, II. slight shivering, III. moderate shivering, IV. intense shivering (91).

cranial internal temperature; this is illustrated in Figure 1-9. Similarly it was shown that variations in skin temperature can modify heat production independent of cranial internal temperature. These observations and many others were summarized by Hammel (82) as is shown in Figure 1-10. In this model a functional set-point is suggested which would serve to regulate temperature. In the case of cold stimulation, the afferent impulses from skin receptors

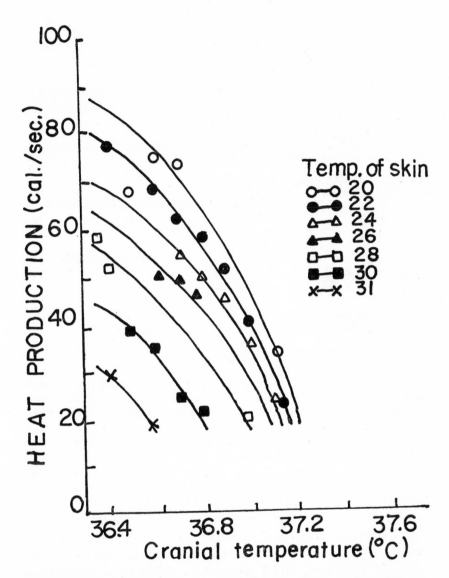

Figure 1-9. Relationship between heat production and internal cranial temperature. For a given skin temperature the heat production is proportional to the central temperature of the brain. The graph also shows that at a constant cranial temperature, heat production is proportional to the temperature of the skin (redrawn from 10).

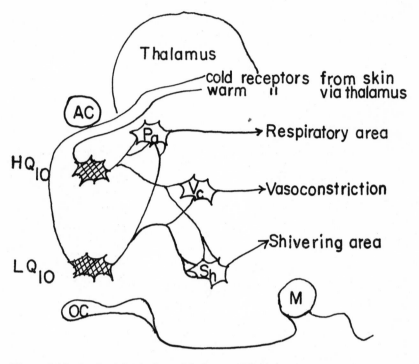

Figure 1-10. A physiological model for establishing a set-point temperature and illustrating the possibilities for adjusting the set point. AC = anterior commissure; OC = optic chiasma; M = mammillary body; Pa = primary neuron for panting; Sh = primary neuron for shivering; Vc = primary neuron for vasoconstriction; cross-hatched cell bodies = low (L) Q_{10} and high (H)Q_{10} primary sensory neurons (82).

would travel via the thalamus to low Q_{10} sensory, neurons and depress respiratory panting area (Pa), while stimulating cutaneous vasoconstrictive (Vc) and shivering (Sh) areas in the hypothalamus. With stimulation of warm receptors, the high Q_{10} neurons would stimulate the Pa area and depress the Vc and Sh areas. The particularly high concentrations of monoamines in different parts of the hypothalamus suggested their importance in the control of the temperature regulation centers (217).

Feldberg and Myers injected minute amounts of monoamines into the third ventricle and in different parts of the hypothalamus (61).

On the basis of these studies, two types of hypothalamic neurons were described: one, sensitive to lowering of temperature, would serve to activate the peripheral heat production systems (serotonin would be the neuroeffector substance implicated in this type of receptor) ; the other, sensitive to increase in temperature, would serve to activate the peripheral heat loss systems (noradrenaline would be the neuroeffector substance implicated in this type of receptor) (62, 167). These studies are very complex and should be interpreted with precaution at this stage. This is illustrated by comparing Figure 1-11, which reports results obtained in monkeys, and Figure 1-12, which illustrates results obtained in sheep (16, 166). In monkeys cold receptors stimulation would release serotonin while in sheep this substance would be released by warm receptor stimulation. Whether these differences are due to species variations or any other factor related to the methodological approach, is still a matter of debate. Recently Myers gave direct evidence, as seen in Figure 1-13, for a substance released into the third ventricle while the temperature of the body is lowered. Indeed cerebro-spinal fluid obtained from the third ventricle of a monkey made slightly hypothermic by exposure in a cold chamber, will produce shivering and marked elevation of body temperature when cross-infused into the third ventricle of another monkey kept at room temperature (165, 168). While this indicates the release of a substance into the ventricle when body temperature is lowered, no direct evidence is available to show that this substance is 5-HT or some monoamine. Reviewing the available evidence on the subject Hellon concluded that "detailed understanding will come when more microintophoresis experiments have been done and when more sensitive assay methods for catecholamines are available" (90) .

INDIVIDUAL FACTORS IN COLD TOLERANCE

The level of endurance to cold varies from one person to the other. This is confirmed not only by subjective evaluation but also by actual measurements. In this case it also becomes more difficult to dissociate the factors involved which determine the degree of discomfort for a given person.

The *size and shape of the body* will affect the amount of heat

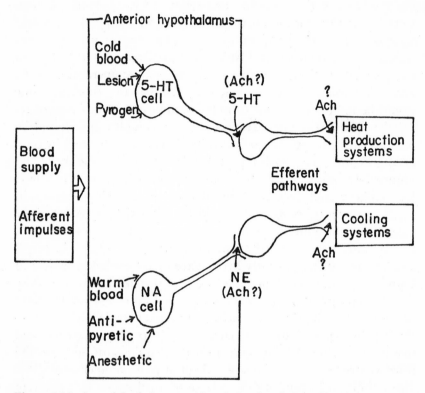

Figure 1-11. A model of the neurochemical role of the cells of the anterior hypothalamus in thermoregulation of the primate. Individual amine-containing cells, 5-HT or catecholamine (NE = norepinephrine and DA = dopamine), are stimulated by efferant impulses or chemical agents. To activate heat production, 5-HT (serotonin) is released in the anterior region and, to activate heat loss, catecholamines are similarly released. Separate efferent pathways delegated to cooling or heating of the primate are envisioned, although there may be some functional overlap in the fiber pathway. Acetylcholine (ACH) may be involved in transmission of the efferent impulses outside of the anterior hypothalamus, but experimental evidence does not suggest its involvement within the anterior region (166).

lost to the environment. For instance, when expressed per unit of body weight, the heat required to maintain a constant internal body temperature will be greater in a child than in an adult. This is because the surface exposed to the environment is greater per

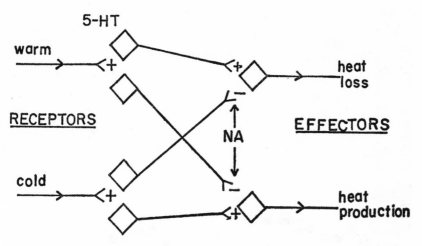

Figure 1-12. A neuronal model based on the effects of intraventricular injections of 5-HT and noradrenaline in sheep under warm and cold conditions (16).

unit of body weight if the total of the body is smaller. For that reason in a given cold environment a smaller individual has to produce relatively more heat than a larger individual to maintain homostasis. All other factors being constant, the shape of the body may also have some importance in cold tolerance. For the same body weight a short person will lose less heat to the environment than a tall one. This is principally because long arms and legs form a larger surface exposed to the environment.

Another important factor in cold tolerance is the layer of *subcutaneous fat*. The thermal conductivity of fat is much smaller than that of muscle. The subcutaneous fat layer is not very well vascularized and consequently its temperature gradient will be affected by environmental temperature; the most important the fat layer, the greater the fall in skin temperature. Figure 1-14 shows that at 70°F, the fat thickness has little effect but at 60° and specially at 50°C, skin temperature drops as subcutaneous fat becomes more important (130). As a result of this, the heat loss to the environment, which depends on the difference between temperature on the surface of the body and that of the environment, becomes much less

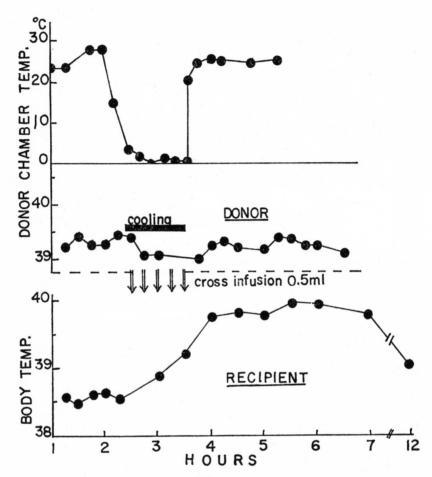

Figure 1-13. After a donor rhesus monkey is cooled by dry ice in its chamber (top tracing), 0.5 ml CSF is obtained from the III^d, ventricle of the donor and cross-infused to the III^d ventricle of a normal recipient monkey. This procedure is repeated at fifteen-minute intervals. Immediately following the first cross-infusion, the normal monkey begins to shiver and its temperature rises (165).

important in fat man. It is for that reason that shivering, as shown in Table 1-I, is much more pronounced in thin persons.

The subject with very little subcutaneous fat (2.25 mm), starts shivering after forty minutes at 60°F and after twenty minutes at

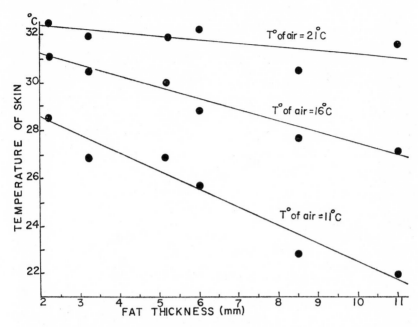

Figure 1-14. Six subjects of different fat thickness were exposed for one hour in the nude at either 11, 16 or 21°C. Average skin temperature changes show an inverse relationship between fat thickness and skin temperature (130).

50°F, whereas the subject with more fat (11 mm) only shivers after sixty minutes at 50°F. When all these results are put together, one can see in Figure 1-15 that a thin man will start shivering at a skin

TABLE 1-I

SHIVERING OBSERVED AT 20, 40, OR 60 MIN. WHEN THE ROOM TEMPERATURE WAS 70°, 60° AND 50° F (130)

Thickness of epimuscular tissues (mm.)	Environmental temperatures		
	70°F	60°F	50°F.
2.25	—	40, 60	20, 40, 60
3.25	—	60	40, 60
5.25	—	—	40, 60
6.00	—	—	40, 60
8.50	—	—	60
11.00	—	—	60

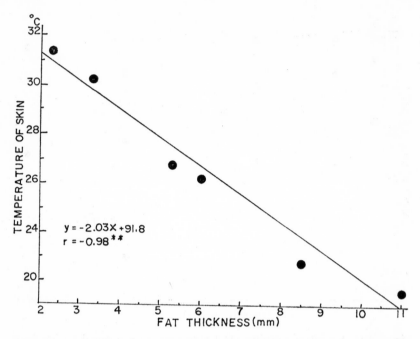

Figure 1-15. Six subjects of different fat thickness were exposed for one hour in the nude at either 11, 16 or 21°C. Skin temperature was noted when shivering was first observed and this is plotted against fat thickness. The figure shows a very significant inverse relationship between fat thickness and the skin temperature at the onset of shivering (130).

temperature of about 87.5°F and that the corresponding temperature for a fat man is around 70°F. It can be concluded from these observations that the subcutaneous fat is an important insulator which may account, for a large part, for the variations in individual cold tolerance. Similar conclusions are reached when fall in rectal temperature or increase in oxygen consumption instead of initiation of shivering are used as the index of cold tolerance (27, 118, 119). Indeed Figure 1-16 shows that thirty-minute immersion in water at 15°C causes a much greater fall in rectal temperature in thin subjects. Similarly Figure 1-17 indicates that the increased heat production due to cold water immersion is much greater in thin subjects because of smaller tissue insulation. These studies would indi-

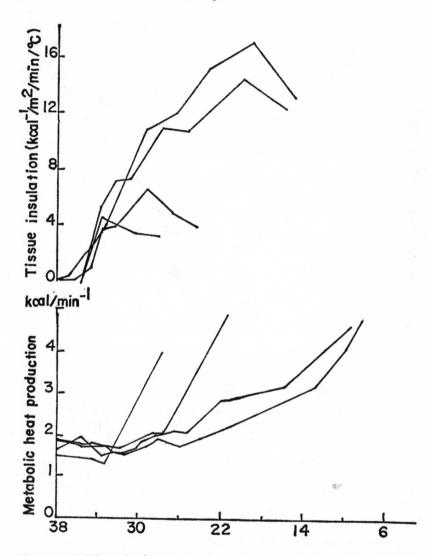

Figure 1-16. Effect of subcutaneous fat on heat production and body insulation in cold water (°C). Results from 2 fat and 2 thin men after they had stabilized their deep body temperatures in stirred water at different temperatures. The 2 lower curves = thin men (mean skinfold thicknesses 6.5 and 6.7 mm). The 2 upper curves = fat men (mean skinfold thicknesses 26.7 and 26.8 mm) (118, 119).

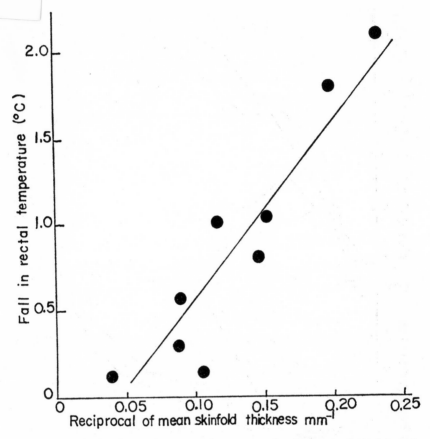

Figure 1-17. Relationship between subcutaneous fat thickness of 10 men and their falls of body temperature during 30 minute immersions in stirred water at 15°C. Skinfold thickness is mean of readings at 4 standard sites (118, 119).

cate that the importance of the subcutaneous fat in cold tolerance is related to its effect on heat conservation. While this is important in all types of cold exposure, mechanisms which accentuate the prevention of heat loss may become more crucial in cold water immersion. Physical activity upon exposure to cold air becomes an important factor in cold tolerance as it increases the heat production. However, in cold water where heat loss by convection is very large, the movements associated with swimming while increasing heat production accentuate the amount of heat loss to the extent that this may become detrimental. Figure 1-18 shows that the fall in rectal temperature is significantly greater in subjects swimming in cold water than in subjects remaining still.

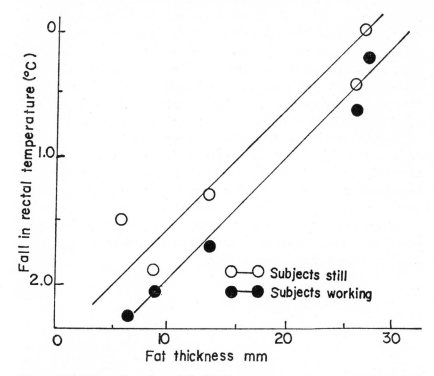

Figure 1-18. Effect of work on fall of body temperature in water just too cold to allow thermal balance (Drawn from table in 118, 119).

Physical fitness may be another important individual factor in cold tolerance. Habitual exercise increases the vascularization and the size of the striated muscles and of the heart, and the working capacity of these tissues is enhanced accordingly. This may have some importance in cold weather and in some conditions it may become the limiting factor. Figure 1-4 will serve to illustrate this. Supposing a subject wearing clothing of approximately one inch in thickness (equivalent to 4 clos) is exposed to a temperature of 10°F, one can see that light work (walking at 3.5 mph) will provide in these conditions sufficient heat to maintain thermal equilibrium. If the same subject is exposed to a sudden fall in environmental temperature and if the wind velocity is increased, the same level of activity will not be adequate to maintain body temperature at a normal level. To illustrate this let us assume that the temperature drops from 10 to 0°F and that the wind goes up to 30 mph, the efficient ambiant temperature will have fallen to —20°F according to Figure 1-5. In order to maintain thermal equilibrium without increasing clothing insulation, the level of activity will have to be three times greater than at rest (running at 7 mph). If these conditions were to prevail for some length of time, the level of physical training could become an important factor in cold tolerance. Indeed it is well known that training can significantly increase the sustained high level of oxygen consumption or the total heat production over a period of time. Physical training may then become an important factor in cold tolerance as one learns to gauge its level of activity with relation to the demands of the environment and at the same time is able to depend on an enhanced capacity for heat production in cases of necessity.

CHAPTER 2

METABOLIC EFFECTS OF COLD

SOURCE OF ENERGY FOR THERMOGENESIS

R ATS LIVING AT 5°C consumed twice as much oxygen as rats kept at 25°C. Similarly the caloric intake of these animals in the cold will double after some time in the cold. In the first few days however food intake is not sufficient to meet the caloric requirements and the animal loses weight; this is reflected by a mobilization of fat reserves. Following this initial phase food intake becomes adequate and the animal assumes normal body growth without drawing upon its energy reserves. During that period the heat production which is supplied at first primarily by shivering, is replaced progressively by nonshivering thermogenesis (146, 210, 88). What substrate is used for heat production in the cold, and is there a difference in that respect between shivering and nonshivering thermogenesis? These questions have been studied and some answers are available. Prior to discussing this point, consideration may be given to the basal metabolic rate of animals exposed to cold. There can be no doubt that the oxygen consumption, measured in a neutral environment and under basal conditions, is significantly increased in animals that have experienced sufficient cold exposure. This increase however is not very large, being on the order of 10 to 15 percent as shown in Figure 2-1. As was indicated already, available evidence points to an increase in thyroid activity compatible with this elevation in basal metabolic rate. However, what significance does this have with regard to cold adaptation? For instance, a very large increase in the basal metabolic rate induced by daily injections of small doses of thyroxine (50 μg/kg) for a week improved cold tolerence, but not as much as cold exposure for the same period of time. Furthermore remarkable degree of adaptation can be induced

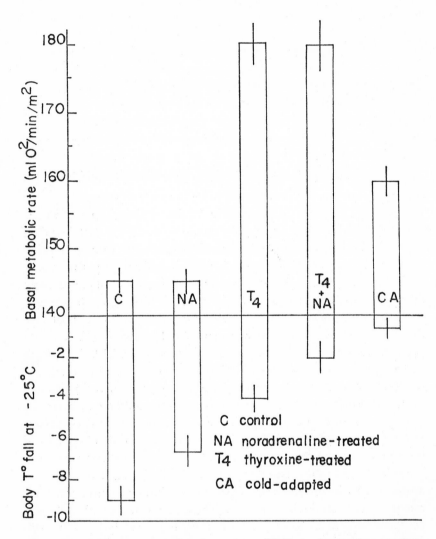

Figure 2-1. Oxygen consumption increase at 26°C one hour after a subcutaneous injection of noradrenaline (30 μg/100 g body weight) in rats treated for 35 days with noradrenaline (30 μg/100 g day), injected with thyroxine (10 μg/kg/day), injected with both noradrenaline and thyroxine each day, and cold-acclimated (35 days continuous exposure at 6°C). The fall in colonic temperature was measured for all groups, three hours after exposure at −25°C (146).

by repeated injections of catecholamines without producing any changes in the basal metabolic rate. Figure 2-1 summarizes this situation and shows that the level of basal metabolic rate cannot be related per se to the level of cold tolerance induced in animals (146). Studies by Depocas indicate that the increased heat production observed in rats exposed to cold is associated with an increase in glucose metabolism. Adaptation to cold, however, did not produce any change in glucose metabolism (47). On the other hand lipid metabolism has been shown to be an important substrate not only for cold resistance but also in cold adaptation. Figure 2-2 shows that upon exposure to cold as heat production is increased, the respiratory quotient goes down indicating an enhanced fat mobilization. However, this supply of energy is a rather complex phenomenon since the respiratory quotient returns to normal values when a steady state metabolism is attained (84).

THE ROLE OF LIPIDS

The increased level of free fatty acids in the blood of rats exposed to 4°C for a few hours is also indicative of some changes is lipid metabolism. Figure 2-3 shows that, as this action takes place, there is evidence of enhanced activity of the hypophyso-adrenal axis (157). That these changes are mediated by an enhanced secretion of catecholamine in the cold (152) is suggested by experiments with reserpine also reported in Figure 2-3. Indeed reserpine when injected causes a massive liberation of catecholamines (197) which indirectly would activate the hypophyso-adrenal axis. This effect of reserpine on blood levels of FFA is not observed in hypophysectomized or adrenalectomized animals. Similarly, a second injection of reserpine at twenty-four hours' interval, fails to elevate FFA in the blood. Since the stocks of available noradrenaline are known to be depressed twenty-four hours after an injection of reserpine, these results taken together indicate that upon exposure to cold, noradrenaline is released and activates the hypophyso-adrenal system which causes lipid mobilization. Figure 2-4 illustrates the obligatory action of noradrenaline in the utilization of lipids upon exposure to cold (158, 71). *In vitro* studies have also shown accelerated lipolysis from epididymal adipose tissue of rats exposed to cold for

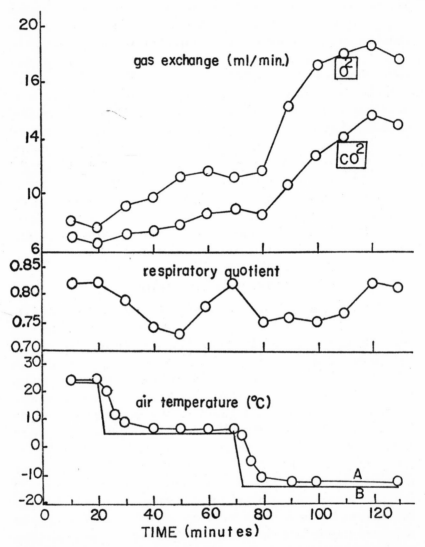

Figure 2-2. Effect of acute cold exposure on respiratory gas exchange of 5 cold-acclimatized rats. Each point indicates the average gas exchange or RQ, continuously monitored with a Beckman $F_{-3}O_2$ analyzer and a Liston-Becker CO_2 analyzer, over the preceding 10-min. period. Vertical lines indicate the standard errors of the mean at the points indicated. In the lower two curves, A indicates the temperature of the effluent air from the metabolism chamber and B indicates the temperature of the environmental test chamber (84).

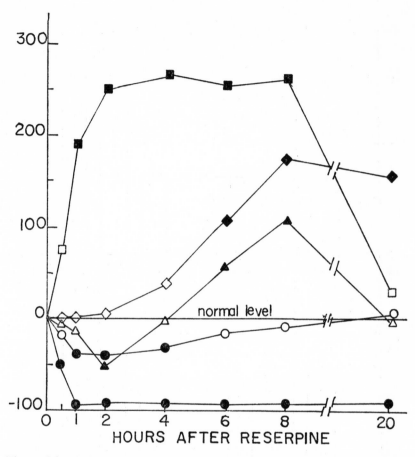

Figure 2-3. Pituitary -adrenal responses of rats to 1 mg/kg of reserpine i.v. Levels of brain 5HT and NE (————) plasma corticosterone (– ⊔ –), plasma FFA (– △ –), adrenal ascorbic acid (– 0 –) and liver TPO (–◇–) are given. Solid symbols represent values differing significantly from normal (P < .05). Each point represents the mean of 10 to 16 animals. Normal values represent the mean of 49 animals (157).

a few hours (159). The results indicate the important role of adipose tissue is supplying fuel for increased heat production in animals exposed to cold. The other question is the importance of lipids in the production of nonshivering thermogenesis which develops with cold adaptation. Experiments in which $C^{14}O_2$ produc-

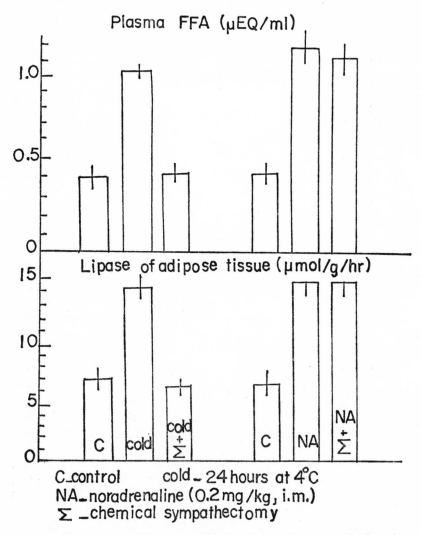

Figures 2-4. Effect of cold exposure (4°C) and noradrenaline injection alone are in association with chemical sympathectomy (produced by giving reserpine in adrenal demedullated rats) on plasma free fatty acid and adipose tissue lipase activity (158).

tion was measured during infusion of palmitate-1-C¹⁴-albumin complex indicate as shown in Figure 2-5 that cold exposure increases

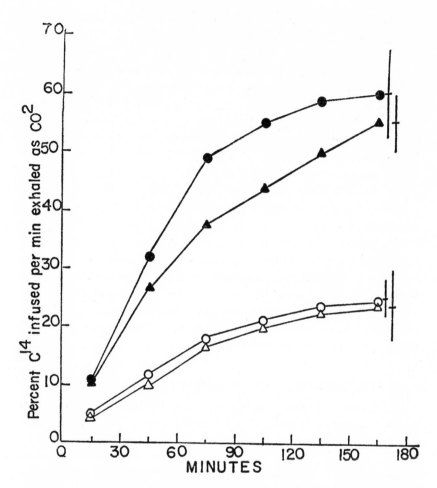

Figure 2-5. Average respiratory $C^{14}O_2$ production during continuous intra-arterial infusion of palmitate-1-C^{14}-albumin complex into non-anesthetized warm— (0, ●) and cold— (□, ■) acclimated white rats at 30°C (open symbols) and 6°C (closed symbols). The results are expressed as the percentage of the activity in the palmitate-1-$1C^{14}$ infused per minute which was recovered in the respiratory CO_1 per minute during successive collection periods of 30 minutes. Vertical lines indicate total range of values for each group of our rats in the last interval of collection of respiratory CO_2 (160).

FFA oxydation, but that cold adaptation does not cause an additional action (160). Lipoprotein lipase activity was shown to decrease in white adipose tissue and to increase significantly in brown adipose tissue, heart and muscles upon exposure to cold (190, 177) ; these changes persisted and were not altered by prolonged exposure. From these results Himms-Hagan concluded that "the triglycerides of the very low density lipoproteins is directed as an energy source to those tissues involved in nonshivering thermogenesis, brown adipose tissue and muscle, and is withheld from the tissue which would normally store the lipid, the white adipose tissue" (110).

It has been shown on the other hand that lipolysis and adenyl cyclase activity of adipocytes from cold-adapted rats are much more stimulated by noradrenaline than similar cells from control animals (209, 210, 211). Thus lipid would seem to be an important fuel for heat production in animals exposed to cold. However there does not seem to be satisfactory explanation for the processes involved in the biochemical control of nonshivering thermogenesis which develops with cold adaptation, although a marked sensitization to the metabolic effects of noradrenaline seems to be an important component of the reactions involved.

THE IMPORTANCE OF BROWN ADIPOSE TISSUE

The role of brown adipose tissue in cold adaptation has been studied by many investigators in recent years. In 1950 Pagé and Babineau (172) reported a marked increase in the size of brown adipose tissue in animals exposed to low temperatures. Since then many workers have been interested in this tissue and have shown its importance in cold adaptation (21, 38, 99, 201, 202, 203, 105, 153, 196). That the brown adipose tissue is related to increased tolerance to cold, is suggested by results of Figure 2-6 showing that the fall in body temperature of animals exposed at $-25°C$ is inversely related to the size of this tissue (146). It is interesting to note that treatments with thyroxine and noradrenaline which produced a hypertrophy of brown adipose tissue without any exposure to cold, were shown to significantly increase cold tolerance. Additional evidence was obtained in which removal of brown adipose tissue in cold-adapted animals significantly reduced both response

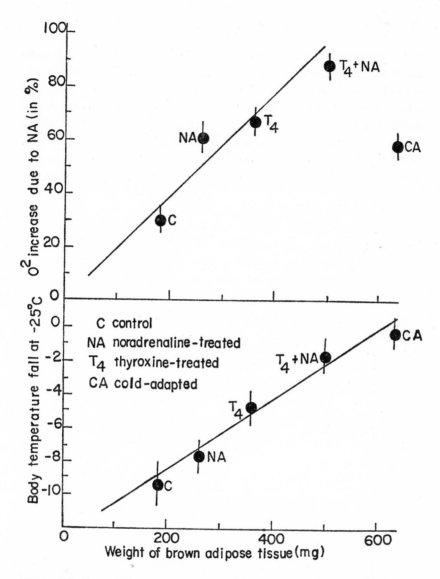

Figure 2-6. Increase in oxygen consumption expressed as percentage above basal values in response to a subcutaneous injection of noradrenaline (300 µg/kg) in control, noradrenaline-treated (300 µg/kg per day for three weeks), thyroxine-treated (50 µg/kg per day for 3 weeks) and cold-adapted rats (6°C for 3 weeks). The lower part of the figure shows the correlation between the size of the interscapular brown adipose tissue and the fall in body temperature following a three hour exposure at −25°C in the different groups studied (146).

to noradrenaline and tolerance to cold (99, 153). Figure 2-7 illustrates these findings (153). Smith and co-workers have concluded, after extensive studies, that the remarkable thermogenic capacity of brown adipose tissue and the very special vascularization arrangements around this tissue contributes to preferential heating of the spinal cord, the heart and the overall thoracic area (201, 202, 203, 105); this effect is certainly not a negligible contribution to an enhanced resistance to cold in cold-adapted animals but it would not

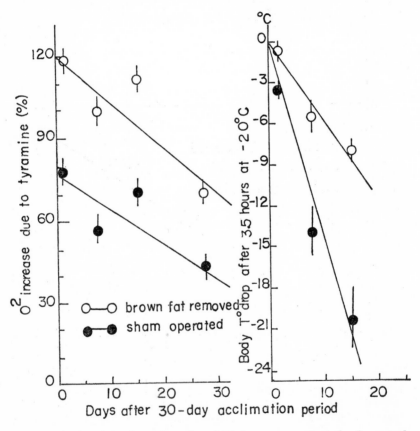

Figure 2-7. Rats were first adapted to 6°C for one month and subsequently the interscapular brown adipose tissue was removed in one group while the other group was sham operated. At different times after this operation, the response to tyromine and tolerance to cold were measured while the animals were kept at 20°C (153).

seem to explain the large participation of this tissue in nonshivering thermogenesis. Another suggestion is that the brown adipose tissue contributes an important source of noradrenaline for the whole of the organism; indeed it has been shown that brown fat of cold-adapted animals contains three times more noradrenaline than nonadapted animals and that the noradrenaline turnover is greatly enhanced by adaptation (38). This extra noradrenaline, if released into the circulation, may well become an important stimulant of thermogenesis, considering the fact that cold-adapted animals become extremely sensitive to noradrenaline. Another suggested role for brown adipose tissue is that of direct control of shivering (21). This control would be exerted by the interscapular brown fat through its special anatomical location, combined with its very high calorigenic properties (21, 89). Figure 2-8 shows that local heating of cervical, but not of lumbar vertebral canal, suppresses shivering and significantly diminishes heat production (20). The role of the brown adipose tissue in cold adaptation and in potentiating noradrenaline has been related to some hormonal factor by Himms-Hagen (99). This suggestion is based on the fact that removal of the interscapular brown fat reduces noradrenaline sensitivity by 40 percent and significantly reduces cold tolerance in cold-adapted animals (99, 153). The logic of this is that the removal of 0.5 g of fat, which is the size of interscapular brown fat in cold-adapted animals, could not explain, on the strict basis of source energy, the results obtained. Thus it seems possible that the brown adipose tissue contributes to nonshivering thermogenesis by releasing into the circulation a hormonal factor such as noradrenaline or some unknown substance.

These results indicate that the brown adipose tissue is important in cold adaptation; much more than the white adipose tissue. However, the mechanisms of its action and its contribution to nonshivering thermogenesis have not been elucidated.

CALORIC REQUIREMENTS IN THE COLD

The caloric requirement as related to environmental temperature is a complex question. Subjective feelings on this matter do not always correspond to reality. For example, urban people not used to

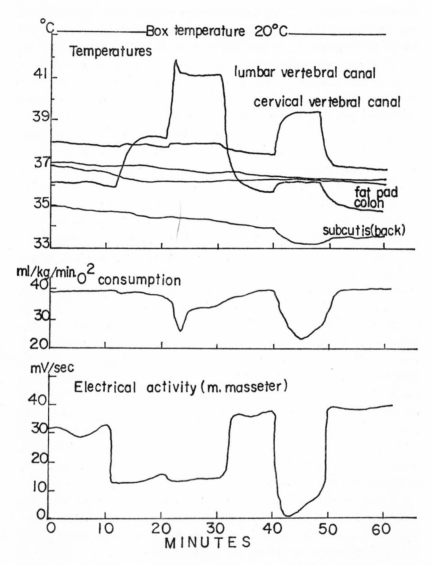

Figure 2-8. Study in a 3-week-old unanesthetized guinea pig, reared at neutral temperature (the extent of nonshivering thermogenesis elicitable thus being low). Shivering as elicited at an environment temperature of 20°C was entirely suppressed by RF heating of the cervical veterbral canal at (C_5-T_1); heating the lumbar region of the vertebral canal was much less effective (20).

manual work may have quite different views on food requirements when compared to rural manual workers. Persons living in a city are familiar with the hot and humid weather and know how to adjust food and especially water intake to cope with the demands of the environment. The same people, when exposed to cold weather on an occasional outdoor excursion, may have a tendency to eat so-called heavy foodstuff while reducing excessively the liquid intake; this behavior frequently leads to stomach and intestinal disturbances. The possible reason for this is that exposure to cool or cold weather often gives a feeling of well being and at the same time a sensation of having used a great deal of calories even when the expenditure has been negligible.

Figure 1-4 may be quite useful in assessing indirectly, but with sufficient relative accuracy, the caloric requirements while exposed to a wide range of temperatures and engaged in a variety of activities with clothing having different insulative values. In this figure, M indicates the metabolism (1M = resting metabolism) and the clo is a unit of insulation (1 clo = approximately $\frac{1}{4}$ inch clothing or a light suit) (25). This figure shows that with a three-inch sleeping bag and an inflated mattress, a person can be comfortable at $-40°F$ if sheltered from wind; in these conditions the caloric requirements need not be different from those required during sleep or rest in a room at 70°F. Similarly, the energy produced during heavy work (M = 6.0) should be sufficient to maintain thermal balance at $-40°F$ even when the garment insulation is limited to 2 clos. These observations stress the value of multi-layer clothing in which a layer can be easily added or taken off in order to adjust to the level of activity and the degree of cold exposure. It has been estimated that Eskimo clothing, which consists of two caribou furs (total thickness of about 3 inches = 12 clos), could supply sufficient insulation under Arctic conditions without increasing the requirements for food intake (185). It is possible for these reasons that the caloric intake of the Eskimos has been estimated by Rodahl as being approximately 3000 Cal/man/day (183).

Caloric requirements have been measured for people living in different climates. Johnson and Kark reported that soldiers consume more food in colder regions (114). These findings were used by Keys who suggested the following formula to estimate the caloric requirements with regard to temperature: Cal/man/day = 5660 — 15.9 T (where T stands for temperature in °F) (120). Subsequent studies did not completely substantiate these conclusions (134). Many types of food surveys have been conducted on military personnel engaged in simulated field conditions. A first method consists in measuring left overs in a standard military ration. This method overestimates the food consumed because "plate waste" is not taken into account. It is also known that certain items such as almonds, raisins, chocolate, etc. are not always returned to the experimenter although they are not consumed. In another method all food intake is accurately measured; this of course in the best method. Finally, collecting "plate waste" has been used as a method of assessing food consumed by large groups of people. This last method and the first one described seem to overestimate food consumption by approximately 10 percent. Taking these factors into consideration, the following estimates of food consumption, which are summarized in Table 2-I, were obtained in some military exercise which took place at different environmental temperatures (134).

As may be seen for a somewhat comparable level of activity, environmental temperature has no effect on food intake in temperate and arctic regions, although some difference may be noticed in the

TABLE 2-I
FOOD CONSUMPTION WITH RELATION TO TEMPERATURE (134)

Food intake Cal/man/day	Environmental temperature (°F)	Location
3400	0	*Shilo, Manitoba Lat. 49°
3600	—12	*Fort Churchill Lat. 57°
4000	—15	*Fort Churchill Lat. 57°
3600	—25	Fort Churchill Lat. 57°
3800	—17	Fort Churchill Lat. 57°
3800	Temperate	U.S.
3100	Tropics	Java

*Located in Canada.

tropics. In view of previous discussion these findings in a sense are not surprising. Assuming a moderate level of activity (3 M) with an Arctic suit which has an insulative value of about 4 clos, a person can withstand temperatures as low as —40 to —60°F without wind and maintain positive heat balance. At night, with the extra insulation provided by sleeping bags, the same extremes of temperatures can be tolerated even when heat production is reduced to basal conditions. Consequently cold per se does not seem to increase significantly the caloric requirements of Arctic populations. Similarly for a given exercise, the energy requirements do not seem to differ greatly with different times of the year. Table 2-II shows that the heart rate increase of soldiers walking at different speeds is comparable in both the winter and summer.

TABLE 2-II

Heart rate variations in subjects walking at different speeds in the winter (at —15°F on hard snow-covered tundra) and in the summer (at 60°F on hummocky dry terrain). The differences between these two conditions are not significant (134).

Speed (mph)	Winter (18)*	Summer (8)*
2.30	102 ± 3**	98 ± 2
2.85	114 ± 3	110 ± 3
3.40	125 ± 3	124 ± 3

*Number of subjects.   **Standard error.

These conclusions are supported by extensive studies done by Welch et al. (219) who found that energy expended, as measured by oxygen consumption, is equivalent to 3200 Cal/man/day for troops performing standard military exercise under Arctic conditions.

Food consumption, however, cannot always be equated with energy expended. This is true, of course, for many world populations with the resulting problems of overweight, especially in the western world. Arctic conditions may tend to amplify this phenomenon. In a group of forty employees of the Canadian Defence Research Board Laboratories, living at Fort Churchill, Manitoba (lat. 57°), an average weight gain of seven pounds was noted over a period of one year (134). These people stayed in well-heated barracks and went outdoors only a few minutes each day during the winter time.

Although some specific conditions, such as the length of the day, may explain these findings, it seems quite likely that other factors, such as isolation or boredom, might be of prime importance. When the level of activity is not imposed by specific tasks, the voluntary rate of energy expended is often dependent on the ability of the body to retain a neutral thermal balance. In cold and temperate climates, the proper use of multi-layer clothing allows a great variety of activities in a wide range of temperatures. In warm climates, however, a neutral thermal balance can only be achieved, in some cases, by reducing the level of activity, thus decreasing caloric requirements. In summary, it would seem that work performed and food intake are somewhat independent of environmental temperature in temperate and cold climates, but that in warm or hot regions food intake may be reduced because of decreased activity. "People seem to adapt to the desert by doing less and less and not by any physiological change within them" (4).

COMMON COLD AND VITAMIN C

Pauling in 1970 published a book on *Vitamin C and the Common Cold* in which the author concludes: "I am convinced that the value of ascorbic acid in providing protection against infectious disease should now be recognized" (173). This statement coming from an eminent scientist, holder of two Nobel Prizes, created some kind of sensation throughout the world. This contributed to the great popularity of this inexpensive and natural food which was already considered by many as an efficient means of preventing and combatting the common cold. Pauling recommends a daily allowance of 250 mgs to 2 grams per day which is quite in excess of the US and British National Research Council recommendations of 20 to 60 mgs. This extra vitamin C would increase blood levels, saturate the tissues and cause many beneficial effects such as improved resistance to stress, increased IQ scoring and possible preventive effect on bladder tumor, atherosclerosis and various infectious diseases. The blood level of ascorbic acid for a normal population is 1.1 mg %; increased intake can easily double this value whereas deficiency results in very low levels of this vitamin in the blood. Values below 0.5 mg % have been shown to be compatible with symptoms of scurvy (1). While scurvy is no more a frequent disease in the western world, the question is raised about the frequency of low levels of ascorbic acid in the blood. In 1954, we had determined the plasma level of this vitamin in 316 persons from a mixed military and civilian northern population living at Fort Churchill, Manitoba. These determinations made during the month of August gave a mean value of 1.1 mg % which was independent of age, except in the age group of fifty to sixty-five where the average was significantly lower than normal at 0.8 ± 0.1 mg % (132) ; this decline with age had been reported previously (121). The same determination was made in March on sixteen subjects who had lived in the North throughout the winter. The level of vitamin C in the blood was found to be 0.7 mg %. While only 6 percent of the subjects had levels lower than 0.5 mg % in August, this value increased to 65 percent in March. Whether this difference is due to low levels of vitamin C in the food towards the end of the winter or to the severe climate cannot be decided from this study. Just the same these observations show the necessity of increasing the intake of this vitamin in northern populations in order to maintain normal blood levels. Of some interest is the fact that in 1954, nine cases of scurvy were treated with success in this area which had a population of about 4000 inhabitants. These were infants, seven from Indian and two from white parents, who had been fed can milk from the time of birth and they were not getting vitamin supplement. Very low levels of vitamin C have been shown to exist in certain conditions. Unless these levels are associated with scurvy, there is no direct evidence that decline of vitamin C in the blood will affect the incidence of specific diseases in humans. On the other hand, many studies have indicated an enhanced resistance to different stresses when supplements of vitamin C are taken (7, 50, 51, 173, 207, 214).

The conclusions of Pauling regarding the usefulness of vitamin C in preventing the common cold are based on many studies (39, 67, 75, 178, 182, 221). Many people objected to Pauling's recommendations, others suggested further studies to reinforce the available positive evidence and a few researchers set out to replicate the investigations done some years ago. Anderson and others in 1972 did a double blind study which included over 800 subjects (6). Half of the subjects received placebo pills while the other half took one

gram of ascorbic acid per day in the form of four tablets for a period lasting from early December to the end of March. Subjects
were also instructed to increase the dose to 4 grams per day for the
first three days of any illness. As shown in Figure 2-9, no effect was
found on the number of colds per subject. Regarding the days of
symptoms per subject the supplement of vitamin C produced an
effect with a level of significance of 96 percent. However the num-

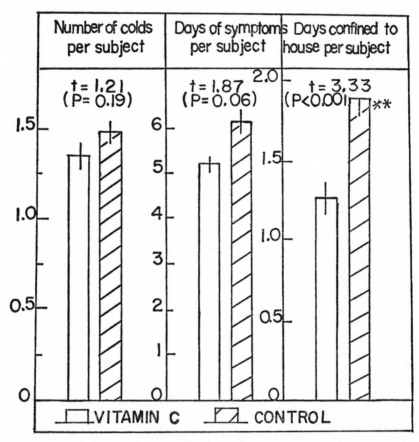

Figure 2-9. Comparison of overall sickness of 407 subjects receiving a supplement of 4 tablets of 250 mg of vitamin C per day and of 411 subjects
given 4 placebo pills for a period of 3 months. This double-blind trial compared the overall sickness experience which is expressed in number per subject. The presence of symptoms and the confinement to the house were also
measured and expressed in number of days per subject (6).

ber of days confined to the house per subject became very significantly (P < 0.001) lower in the group receiving vitamin C. The authors concluded that "the reduction of disability appeared to be due to a lower incidence of constitutional symptoms such as chills and severe malaise, and was seen in all types of acute illness, including those which did not involve the upper respiratory type." A similar study done on a somewhat smaller group gave essentially

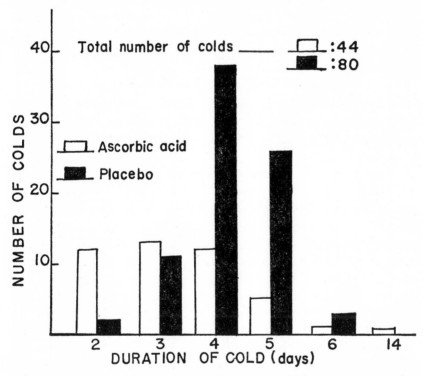

Figure 2-10. Comparison of total number of colds and of duration of cold symptoms in 47 subject receiving 1 g. ascorbic acid per day for 4 winter months and in 43 control subjects given placebo pills for the same duration (32).

similar results. The experimental group (N = 47) received 1 g. ascorbic acid per day between November and March, while the control group (N = 43) was given placebo. Figure 2-10 shows a marked reduction (49%) in the incidence of cold in the ascorbic acid group. The same figure shows that this reduction is due to a significant decrease in the number of colds of four- and five-day duration.

These studies as well as those cited by Pauling indicate some beneficial effect of vitamin C in the prevention and cure of common cold. However, in view of the large doses prescribed and because of the complexity of this problem or of any nutritional studies of this nature, one wonders whether it is justified at this stage to be extremely categorical on the opportunity of taking large supplements of vitamin C for the purpose of fighting the common cold.

COLD EXPOSURE AND VITAMIN C

The role of vitamin C on resistance to cold exposure has also been investigated. Dugal and co-workers (45, 53, 54, 212) did some experiments on monkeys, a species which like humans does not synthetize vitamin C.

Two types of experiments were conducted; one in which the animals were previously exposed to cold while receiving supplements of either 25 or 325 mgs of vitamin C per day. Figure 2-11 shows that higher doses of vitamin C maintained higher rectal and muscle temperature in the animals exposed to cold and significantly reduced the incidence of frostbite. In another experiment, the effect of vitamin C was tested on monkeys kept at room temperature during the treatment and subsequently exposed to −20°C to estimate the degree of tolerance to cold. The larger dose of vitamin C proved effective in delaying hypothermia, as seen in Figure 2-12 but the incidence of frostbite was not decreased in this experiment. These studies show the beneficial effect of vitamin C in cold resistance and also indicate that this effect is maximum when vitamin C supplements are given while the animals are simultaneously exposed to cold. More recent experiments confirmed these findings and have indicated an inadequate synthesis of ascorbic acid in rats fed a low-protein diet (9).

Following these studies on laboratory animals the effect of vitamain C supplement on resistance to cold in humans was also tried (131). In this experiment twelve subjects, wearing shorts and socks only, were exposed for eleven days at a temperature of 15°C while

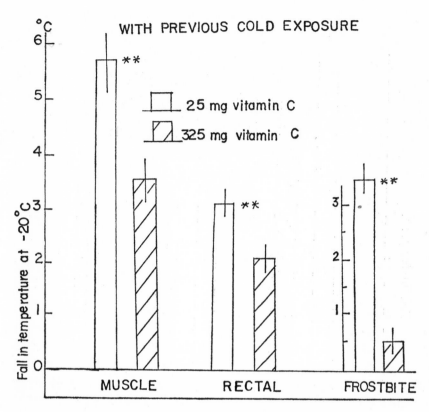

Figure 2-11. Effects of exposure to cold (−20°C for 3 hours) on muscle and rectal temperatures, and incidence of frostbite in monkeys receiving either 25 or 325 mg of vitamin C per day for 6 months while kept continuously at 10°C (212).

being fed a low calorie diet composed of jelly candies giving a total of 600 calories per day. Six of these subjects received 25 mgms vitamin C per day, the other six 525 mgms. In the high vitamin group, skin temperature remained higher than in the other group as seen in Figure 2-13. At the end of the experiment all subjects complained of some foot trouble which was probably related to the combination of cold exposure, low calorie diet composed exclusively of carbohydrate and the relative lack of activity. Table 2-III shows that the larger intake of vitamin C greatly attenuated the symptoms and the duration of the foot troubles that were experienced by these subjects.

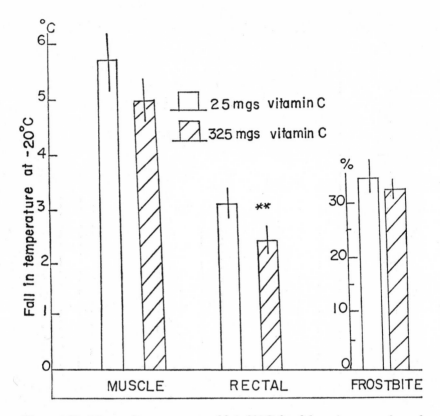

Figure 2-12. Effects of exposure to cold (−20°C for 3 hours) on muscle and rectal temperatures, and incidence of frostbite in monkeys receiving either 25 or 325 mg of vitamin C per day for 6 months while kept continuously at 10°C (212).

On the basis of these studies as well as many other reported by Pauling, the Canadian and American rations issued to military personnel in the North are now being supplemented with approximately 500 mgms vitamin C per day.

In view of the significant evidence accumulated on the beneficial effects of vitamin C, it must be concluded that other studies on the role of this vitamin in the resistance to different stressing conditions would certainly be justified.

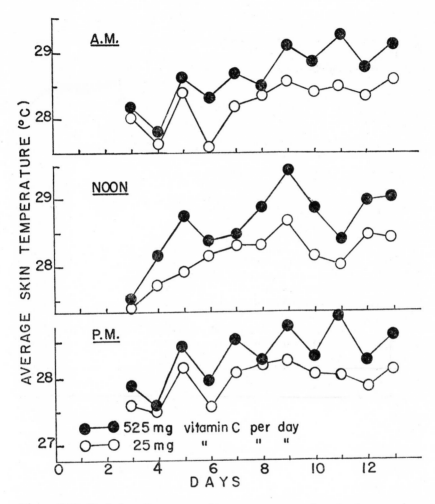

Figure 2-13. Variations in average skin temperatures taken in the morning, at noon and late afternoon, in two groups of human subjects kept at 60°F for 11 days and wearing shorts and socks. One group was receiving 25 mg of vitamin C per day while the other group received 525 mg (131).

TABLE 2-III

Effect of vitamin C supplement on incidence of foot trouble observed in subjects exposed for 13 days at 60°F while wearing shorts and socks only and being fed 600 calories per day (131).

Subject	Description of foot trouble	Duration of foot trouble (days)
	Vitamin C intake 525 mgm per day	
J.B.	Some muscular weakness	2
F.J.	Slight burning sensation	2-3
A.G.	Sensation of pins and needles	2
B.C.	Sensation of pins and needles	2
W.B.	Sensation of pins and needles	1
R.J.	Soreness of the soles with some pain when walking	7
	Vitamin C intake 25 mgm. per day	
M.B.	Severe aching, which would occasionally keep him awake at night with extreme difficulty in walking	15
R.W.	Noticeable swelling of feet with difficulty in walking	10
S.P.	Swelling of feet with difficulty in walking	10
L.W.	Some muscular weakness with occasional pain	8
A.G.	Frequent pains and soreness	10
G.H.	Sensation of pins and needles	3

CHAPTER 3

ENDOCRINES

M ANY FEATURES SEEN in some animal species seem to have been caused by long-term exposure to a given environment. Length of appendages, growth of hair, thickness of insulation, and rate of sweating are among the many specific features which have been influenced by the temperature of different parts of the globe. For many species living in arctic or desert conditions, these features, added to special behavioral habits, have been important for protection from the severity of the climate. The arctic fox, for example, can withstand temperatures as low as $-30°C$ without increasing its metabolism (194). There are many species, however, which depend on physiological responses for surviving in their environment. For instance many domestic and wild animals, as well as humans, are often exposed to a wide range of temperatures for which they are not readily adapted. However, these same animals show a remarkable capacity for adapting to a new environmental temperature through an improved capacity for metabolic heat production. The prominent role of catecholamines in that respect has been discussed previously; other endocrines have also been shown to be implicated in the processes of resistance and adaptation to cold.

THE THYROID GLAND

The current opinion has favored for many years the participation of the thyroid gland in cold adaptation. Following earlier studies, it was shown that the conversion ratio of I^{131} (PBI I^{131}/total I^{131}) was much higher in animals exposed to cold for eight to sixty days than in animals maintained in a neutral environment (36). An increased turnover of thyroid hormones was also reported during cold exposure (36). Another approach which gave interesting results consisted in assessing the amount of thyroxine required to prevent thyroid hypertrophy in animals exposed to cold. The quantity of hormone needed was found to be twice as large at 5°C than at room temperature (223). On the other hand, it has been reported that the histology of the thyroid and the uptake of iodine return to normal within a few weeks after removal from the cold (204, 151).

More recently studies have been reported on kinetics of thyroxine in rats exposed to cold (Fig. 3-1). While plasma thyroxine is lower in cold-adapted animals (3.55 ± 0.33 μg/100 ml plasma) than in controls (5.43 ± 0.35), the secretion rate of this hormone is comparable in both groups (0.22 ± 0.01 μg/hr and 0.21 ± 0.01) because of a larger metabolic clearance rate after adaptation to cold (3.08 ± 0.05 ml/hr) than before (2.36 ± 0.06 ml/hr) (113). Another recent study reported a threefold increase in iodine uptake during cold exposure, and since the iodine concentration in the gland remains normal, these results are taken as evidence for an increase in thyroid hormone secretion in the cold. On the other hand, the same study shows that the increased thyroxine turnover in the cold could simply be due to an enhanced clearance of this hormone from plasma through gastrointestinal pathways (122). The logical conclusion is that the amount of thyroxine available to peripheral tissues during cold adaptation is within normal range (69). Similarly when thyroid function was estimated by the method of isotopic equilibrium while controlling dietary iodine intake, it was reported that cold adaptation did not increase thyroid activity (26).

Recent observations by Héroux and Petrovic (94) are certainly relevant to the present discussion. These authors have shown that animals fed a low bulk thyroxine-free diet, supplemented with KI in the drinking water, showed no hypertrophy of the thyroid, no increase in BMR and no increase in thyroxine turnover rate upon exposure to cold; yet these animals developed a much higher degree of cold tolerance than animals fed the usual commerical chow diet (5). These recent results as well as some from older studies would indicate a limited participation of thyroxine in cold adaptation.

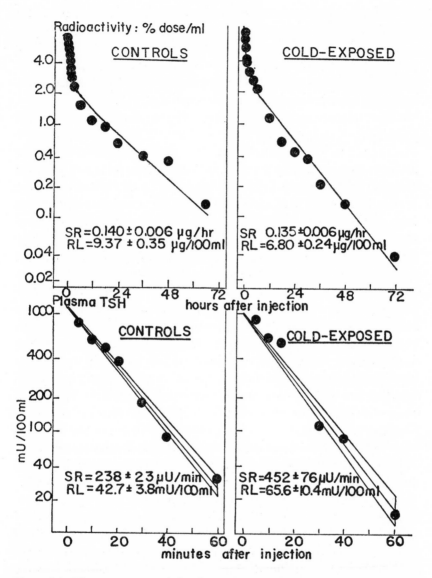

Figure 3-1. The upper part of the figure shows the log of plasmatic concentration of I[131] (in % of injected dose) measured in function of time after injection of thyroxine-I[131] in controls and cold-adapted rats. In the lower part of figure are given the concentrations of exogeneous TSH measured after the injection of 100m U in function of time after injection. SR (the secretion rate) and RL (the resting level) are shown in all cases (55).

These conclusions may hold for thyroxine which is only one form of iodinized hormone secreted by the thyroid. Similar studies on triodothyroxine might prove interesting because of the different properties of this hormone and in view of the fact that a recent study has shown a decrease in the binding of labelled 1-triodothyroxine to serum proteins of cold-acclimated rats (37).

The importance of the thyroid in the development of cold adaptation is still a matter for discussion. It must be agreed, however, that while the methods used to assess secretion of thyroxine are adequate, it is possible that they do not depict small minimal changes. Because of the high calorigenic action of thyroid hormones and in view of the interactions that they have with catecholamines and corticosteroids, it is not impossible to think that minimal changes in thyroxine secretion could be sufficient to exert a somewhat important role in cold adaptation. Of course, it must be kept in mind that cold adaptation is a rather complex phenomenon. While the initial shivering response is replaced by nonshivering thermogenesis, at the same time a given temperature becomes less and less imposing with length of exposure; indeed the heat loss decreases as the insulation increases and as the animal gains weight, thus reducing the ratio between body surface and body weight. For these reasons the importance of thyroid hormones might vary depending on the site of action of thyroxine, the degree of interactions with other hormones, and the duration of cold exposure.

Another approach which has been used to study the relationship between thyroid activity and cold adaptation has consisted of injecting thyroxine daily for a few weeks while the animals are kept at a neutral temperature; at the end of cold exposure, the degree of adaptation is evaluated by measuring body temperature changes in severe cold (191). This method of assessment proved useful and was reevaluated by us recently (146, 216). Groups of rats were injected daily for thirty-five days with either thyroxine (50 mg/day), noradrenaline (300 mg/kg/day) or a combination of the two. These animals along with control and thyroidectomized animals were kept at 26°C for the duration of the treatment. Following this period, metabolic responses to noradrenaline, to thyroxine and to noradrenaline and thyroxine were measured. As shown in Figure 3-2, response to noradrenaline is enhanced in noradrenaline

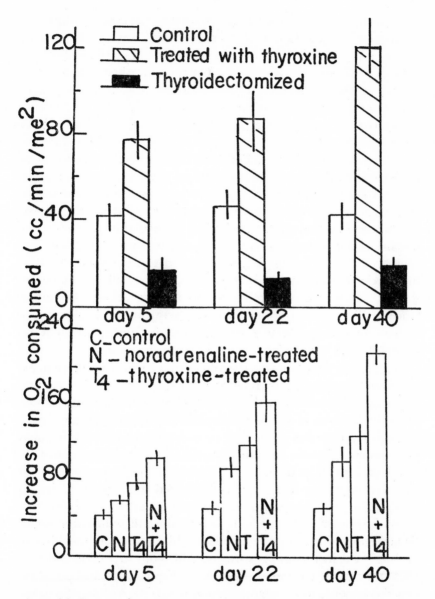

Figure 3-2. Increase in oxygen consumption above basal values in response to noradrenaline (300 μkg) injected subcutaneously at various times (5, 22, N.B. and 40 days) in different groups of rats: controls, injected daily with thyroxine (50 μg/rat), injected daily with noradrenaline (300 μg/kg), injected daily with thyroxine and noradrenaline and thyroidectomized (146, 216).

or thyroxine-treated animals and it becomes significantly reduced in thyroidectomized animals. Combined treatment with noradrenaline and thyroxine resulted in a simple summation of action without evidence of potentiation. On the other hand, response to thyroxine was unchanged in all conditions (Fig. 3-3) in spite of significant changes in basal metabolic rates (Fig. 3-4). Confirming common observations, these results show that the basal metabolic rate is greatly affected by the state of thyroid activity but that the sensitivity to thyroxine is independent of the level of secretion of this hormone. Response to noradrenaline however, depends on normal thyroxine secretion; it may also be increased in the presence of larger amounts of thyroxine, but it is greatly reduced when thyroxine secretion is abolished. Considering metabolic heat production, noradrenaline is more important than thyroxine. The role of thyroxine in that respect is primarily related to its supporting effect for maximal response to noradrenaline, and an alleged enhanced thyroid activity has been reported in cold adaptation. The importance of these activities was tested by measuring cold tolerance of animals treated in the manner described for the experiment reported above.

Figure 3-5 shows improved cold tolerance in the different experimental groups and by referring to Figure 3-2, one can see that the groups which show the best resistance at $-25°C$ are those which respond the most to the metabolic effect of noradrenaline; this is particularly true for the group treated with both thyroxine and noradrenaline in which maximal response to noradrenaline was observed and which, at the same time, has developed tolerance to cold almost comparable to that of cold-adapted animals ($6°C$ for 35 days) (146). At this point it is necessary to restrict somewhat this otherwise appealing Cartesian reasoning since these findings cannot be altogether simply equated. Indeed while resistance to cold is almost as good in the animals treated with both thyroxine and noradrenaline as in cold-adapted animals, Figures, 3-2 and 3-5 show that the response to noradrenaline following this treatment is much greater than that observed in cold adaptation (146). Obviously the role of noradrenaline in cold adaptation is important but more work is needed before it is completely understood.

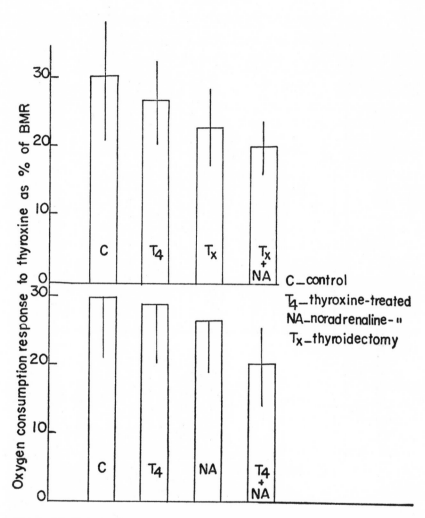

Figure 3-3. Increase in oxygen consumption as percentage of basal metabolic rates in response to thyroxine (75 μg/100g) in different groups of rats: controls, injected daily with thyroxine (50 μg/rat), injected daily with noradrenaline (300 μg/kg) and injected daily with both tyroxine and noradrenaline for 40 days in all groups. Similar responses were measured in thyroidectomized animals and in animals thyroidectomized and subsequently injected for 40 days with noradrenaline (300 μg/kg) (216).

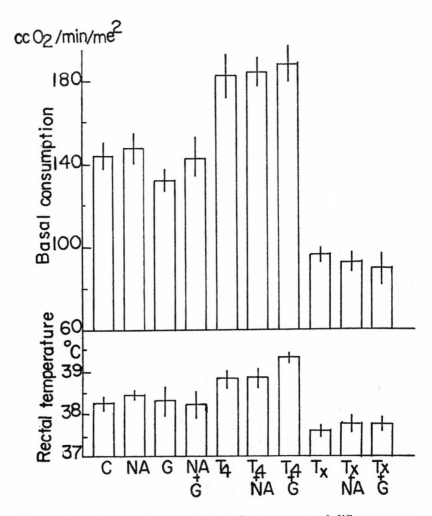

Figure 3-4. Basal metabolic rate and rectal temperature of different groups of rats at the end of a treatment lasting 40 days: controls, injected daily with noradrenaline (300 μg/kg), injected daily with thyroxine (50 μg/rat), injected daily with guanethidine (10 mg/kg) injected with both noradrenaline and thyroxine, or with both noradrenaline and guanethidine, or with both thyroxine and guanethidine, thyroidectomized, thyroidectomized and injected for 40 days with either noradrenaline or guanethidine (216).

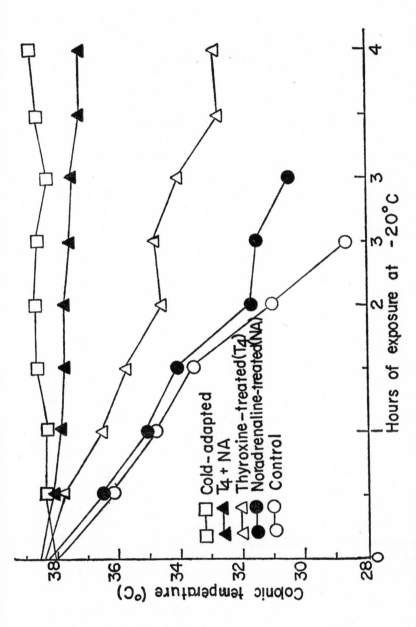

Figure 3-5. Colonic temperature variations in various groups of rats exposed for 4 hours at −20°C. Control animals were compared animals injected daily for 35 days with: thyroxine (50 µg/rat), or noradrenaline (300 µg/kg), or with both thyroxine and noradrenaline (146).

THE ADRENAL CORTEX

A variety of stress conditions produce a nonspecific activation of many endocrine systems. In that respect the original observations by Cannon for the autonomic nervous system (28) and by Selye for the adrenal cortex (192) have been verified many times. Prolonged exposure to heat and cold produce an increased secretion of ACTH, a release of adrenal corticoids, a fall in adrenal ascorbic acid and cholesterol, an adrenal hypertrophy, and an activation of the sympathetic nervous system as evidenced by a marked increase in adrenaline and noradrenaline secretion. At the early stage of cold exposure, that is within one hour, there is a marked increase in plasma corticosterone (18, 112, 113) which persists for a few days. This initial response, which was described as nonspecific by Selye, is possibly of nervous origin since it is abolished by section of pituitary stalk (64). The activity of the adrenal cortex during the first few days of exposure to cold reflects the severity of the stress and is part of the resistance mechanisms. On the other hand following this period of adjustment to stress, many of the activities taking place at the first line of defense cease if the new environment is compatible with survival. Thereafter the organism will rely on resources which normally do not come into play, and the induction of these new mechanisms will confer on the animal an improved resistance which may be called adaptation. This induction takes some time to develop and during this period temporary mechanisms permit resistance. There cannot be any doubt as to the importance of noradrenaline in cold adaptation. The induced metabolic sensitivity observed when the secretion of noradrenaline is enhanced by cold exposure is directly related to an improved tolerance to low temperature. Figure 3-6 gives a schematic representation of endocrine functions in the process of cold adaptation.

At the early stage of cold exposure the nervous and physical responses are probably maximal. The psychological effect of the stress, the disturbance of shivering, the painful sensation of cold and the additional requirement for heat production explain the marked increase in endocrine stimulation, the loss of body weight and possibly some degree of sleep deprivation. The requirements for heat would be met at this early stage mainly by muscular ac-

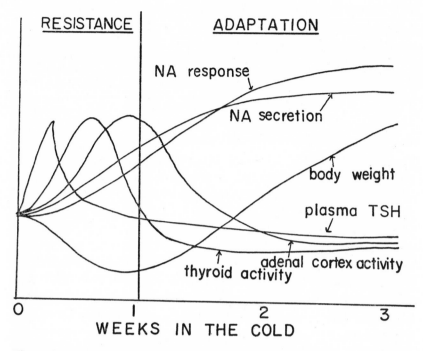

Figure 3-6. Schematic presentation of some variables in rats exposed to cold (5°C). During the first week of exposure the thyroid, the adrenal cortex and the noradrenaline secretion are greatly stimulated. Thereafter the thyroid and adrenal activities return to normal levels, while the secretion and response to noradrenaline remain elevated. Body weight falls during the first week because of insufficient food intake but the growth thereafter assumes a more or less normal pattern.

tivity and to a lesser extent by chemical processes. This nonshivering thermogenesis could be due to gluconeogenesis because of increased glucocorticoid secretion and also to preferential carbohydrate mobilization through an increased adrenaline discharge. If cold exposure is continued, noradrenaline secretion remains elevated and, as shown in Chapter 2, the metabolic effect of this amine becomes very important. At this time, adrenaline, corticosterone and thyroxine secretion have returned to normal levels, shivering has ceased, the animal grows normally and cold tolerance is increased. It would seem that the importance of noradrenaline in cold adaptation is related to a greater capacity for this hormone to mobilize as well as to oxidize nonesterified free fatty acids (83). The adrenal cortex and the thyroid would then have a supportive role but would not be as directly involved as noradrenaline in the control and maintenance of adaptation.

THE PITUITARY GLAND

Fortier and co-workers have recently established pertinent facts about pituitary secretions in cold-exposed animals (65). The secretion rate of TSH is increased in cold-adapted rats as shown in Figure 3-7 (113). This finding is consistent with the current belief of a feedback control of TSH secretion by levels of plasma thyroxine; indeed Figure 3-1 shows a marked fall in plasma thyroxine of cold-adapted animals concomitant with increased TSH secretion. The same group has shown that nociceptive conditions which stimulate ACTH release simultaneously inhibit TSH secretion with the exception of exposure to cold which enhances both TSH and ACTH secretion. This finding therefore fails to support the assumption that enhanced ACTH secretion is incompatible with concomitant increased TSH secretion (113). Another interesting observation is that the enhanced TSH secretion observed during cold exposure can be completely prevented by minor environmental disturbance, such as the presence of an observer, whereas the ACTH secretion is increased in the same situation (55). Furthermore, concurrent hippocampus stimulation suppresses the effect of this emotional disturbance and restores normal TSH response to cold. These findings are summarized in Figure 3-8 (65). These findings would indicate that upon initial exposure to cold a specific TSH secretion and thyroid activation would take place, possibly because of hippocampal stimulation; if exposure is prolonged, enhanced TSH secretion could be explained by some feedback mechanisms activated by the low levels of thyroxine in the plasma. The secretion of ACTH, as estimated by plasma corticosterone concentration, would increase very rapidly in the cold and would result from a nonspecific action possibly of nervous origin. Contrary to what is observed with TSH, the level of secretion of ACTH would then oscillate within normal ranges following this initial and temporary increase.

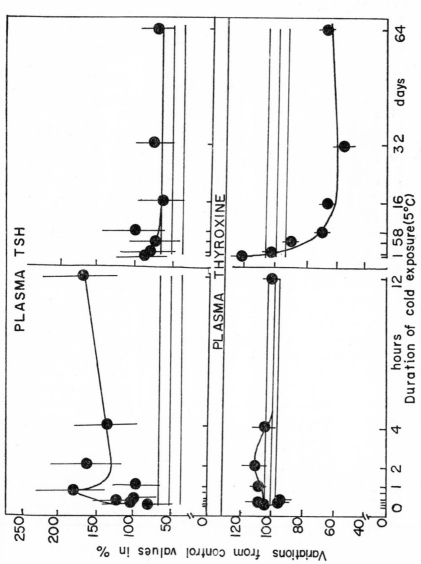

Figure 3-7. Plasma TSH and thyroxine in rats kept at 5°C for 64 days. Results are expressed as percentage of control values (113).

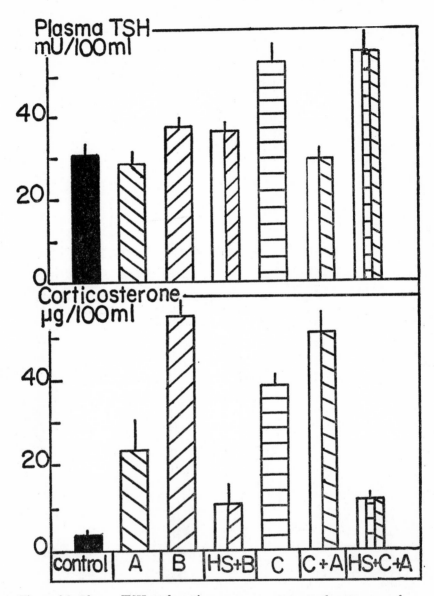

Figure 3-8. Plasma TSH and corticosterone responses to the presence of an observer in the experimental room (A), nicking of the tail (B) in association or not with electrical stimulation of the dorsal hippocampus (HS), and exposure at −5°C for 20 minutes (C) associated or not with the presence of

CHAPTER 4

CARDIOVASCULAR RESPONSES

WHEN THE VARIATIONS in environmental temperature become more demanding, some regulatory mechanisms change progressively and become more important. If the air temperature becomes warmer, cutaneous vasodilatation and increased blood flow will favor an increased heat loss through conduction and convection while enhanced sweat gland activity will contribute to lowering body temperature by evaporation. The degree of tolerance to heat which eventually becomes a function of sweating rate, is accordingly controlled by the capacity for heat loss rather than a decrease in heat production. On the other hand, when the air temperature becomes cooler, cutaneous vasoconstriction will prevent excessive heat loss and on many occasions this may prove sufficient to maintain homeothermia; however, if the environment becomes colder, extra heat production through shivering is the normal protective mechanism. Consequently the degree of tolerance to cold depends initially on heat preservation through a decrease in cutaneous circulation which minimizes heat loss, but subsequently the primary factor becomes an increase in heat production.

Our knowledge of the role of endocrines, substrate utilization or the importance of catecholamines, to name a few of the factors concerned with heat production, has evolved from studies per-

an observer and with or without concomitant stimulation of the hippocampus. The animals were killed 20 minutes after the onset of this procedure. Hippocampal stimulation (18) resulted in nearly complete suppression of the plasma corticosterone rise excited by nicking of the tail or by exposure to cold in the presence of an observer, and (26) fully restored by plasma TSH response to cold, otherwise prevented by the associated disturbance (55).

formed almost exclusively on laboratory animals. The reason for this is very simple: except for a few primitive populations such as the Australian aborigines (79), human beings are seldom exposed to prolonged severe cold conditions which could produce shivering and hormonal activation necessary for the development of nonshivering thermogenesis.

This does not imply that nonshivering thermogenesis cannot take place in humans; it has indeed been observed under special conditions in adults (41) as well as in newborn infants (43, 163). Another fallacy would be to consider the Eskimos to be an ideal population for studies on metabolic adaptation. The Eskimos have learned to live in the Arctic; they have developed protective clothing so efficient that they are very seldom exposed to temperatures normally conducive to shivering and cold discomfort (205). The bulky clothing that they wear, while providing an effective insulation in the cold, might even produce reactions typical of heat exposure especially when the level of physical activity is increased during the milder season. This, of course, does not become critical for the Eskimos whose primary concern remains protection against cold, but it may explain the unexpected findings of a study showing a greater number of sweat glands in Eskimos than in Caucasian or even Negro subjects (117). For all these reasons metabolic responses to cold have been studied primarily in laboratory or wild animals.

CUTANEOUS CIRCULATION IN THE COLD

On the other hand, while the occurrence of whole body cooling and shivering is limited even in Arctic conditions, it is quite common to experience varying degrees of cooling and freezing of the extremities and face even in moderate climates. For these reasons cardiovascular responses to local cooling of extremities have been extensively studied in humans, while less research was done on laboratory animals. Lewis in 1930 was the first to study the effect of temperature on hand blood flow (154). He observed a rapid fall in finger temperature when the hand was immersed in ice water. The finger temperature fell to almost 0°C, but after a few minutes it rose by some 5 to 6° and fluctuated thereafter between 0 and 5°C. This temperature increase caused by an increased blood flow was termed the "hunting" phenomenon (154). Further studies were made showing that this cold-induced vasodilatation was due to an increased blood flow resulting from the sudden opening of the arteriovenous anastomoses (77, 25). This in turn produces marked variations in skin temperature as illustrated in Figure 4-1. One can also see that the pain sensation increases with falling skin temperature and that the rewarming of the skin temperature (from near 0° to 4-5°C) by the so-called cold-induced vasodilatation, completely eliminates this sensation giving the impression of having the hand immersed in lukewarm water. Reappearance of vasoconstriction coincides with an elevation of pain sensation. Frequently rewarming the hand at room temperature produces a marked pain sensation. This increase in pain sensation is shown to take place at a time when skin temperature goes up very rapidly. Consequently pain sensation is observed not only when the skin is cooled but also when it warms up after cooling. Consequently, cold pain is not directly related to the absolute drop in skin temperature. In Chapter 1, it was also shown that shivering is not always directly related to absolute changes in skin temperature. Wolf and Hardy similarly reported that if the hand is immersed in a water bath where the water is cooled from 20° to 0°C by successive steps over a period of one hour, no pain was felt (222). The evidence available would then seem to relate cold pain not so much to absolute skin temperature but to rate of change in the thermal gradient of the hand.

SYSTEMIC CARDIOVASCULAR RESPONSES IN THE COLD

The cold pain induced by immersion of the hand in cold water coincides very closely with a systemic blood pressure increase. This response is the basis of the cold pressor test suggested by Hines and Brown for detecting potential hypertensives (101). While this test has been studied extensively in that respect, it has been generally easier to classify individuals as hypo-, moderate or hyperreactors to a cold stimulus (222). Cold pain caused by local exposure produces systemic responses all indicative of an activation of the sympathetic nervous system. Indeed as shown in Figure 4-2, concommitant with the pain sensation, many signs of increased sympathetic

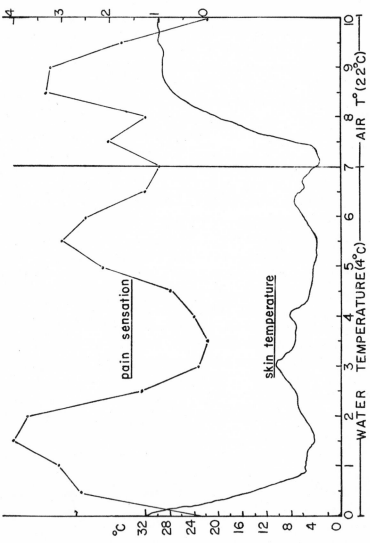

Figure 4-1. Variations in skin temperature of the finger and of pain sensation when the hand is immersed in water at 4°C. Skin temperatures variations reflect constriction and dilatation of the vessels; the diminution of pain coincide with the episodes of vasodilation (unpublished results).

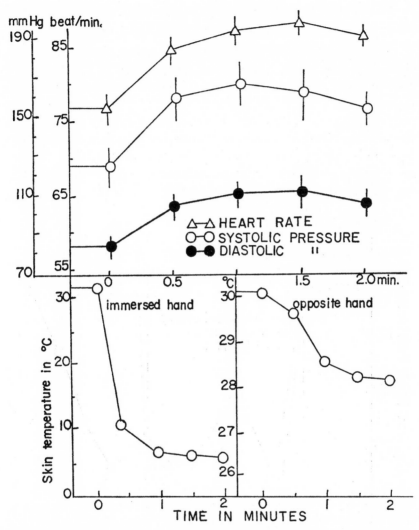

Figure 4-2. Blood pressure, heart rate and skin temperature responses to the immersion of one hand into cold water (4°C) for two minutes (140).

activity are observed, such as elevation of both systolic and diastolic pressures, increase in heart rate and systemic vasoconstriction as suggested by the fall in skin temperature in the immersed hand as well as in the opposite hand.

EFFECTS OF COOLING FACE AND HANDS
ON SYSTOLIC TIME INTERVALS (STI)

These reactions are typical for the extremities, but exposure of the face to cold produces special responses which are interesting to analyze. Early observations reported a marked bradycardia in the diving seal in which heart rate goes from 100 to a frequency of 10 after submersion, while central blood pressure is maintained within normal rage (107) as shown in Figure 4-3. This is made possible through a shift in circulation by which peripheral blood flow is greatly reduced in order to retain adequate circulation in brain and heart tissues in spite of a reduced cardiac action. This remarkable

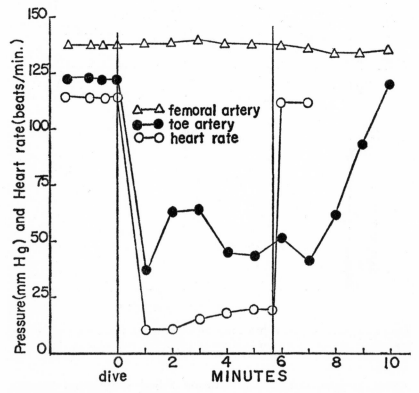

Figure 4-3. Variations in femoral and toe artery pressure, and in heart rate of a seal during dives. The end of the dive in each measurement is marked by a short vertical line (107).

adjustment has also been observed in humans (107) and more recent studies have shown the complexity of this reaction (128). These unpublished data will be given in more detail. A group of twenty-five normal male students ranging in age from twenty-five to thirty were subjected to the following procedure. After resting for about thirty minutes in the afternoon, control measurements were made of blood pressure, heart rate, respiration and systolic time intervals (STI). These determinations were continued during the hand immersion test which consisted of immersing the hand to the wrist in water at 4°C for a period of two minutes. After a rest of forty-five minutes, the same measurements were made during the face immersion test which consisted of immersing the face in water at 4°C for two minutes while the subjects were breathing freely through a snorkle. This procedure was adopted after preliminary studies had shown that none of the variables measured were affected when either the face or hand immersion test was performed with water at 30°C. The measurements of STI were made from simultaneous recordings of electrocardiogram (lead II), phonocardiogram, and carotid arterial pulse tracings on a multichannel photographic recorder operated at paper speed of 100 mm/sec. Respiration was also recorded with a chest pneumograph. From these recordings illustrated in Figure 4-4, heart rate and the different STI were computed. The left ventricular ejection time (LVET) was measured from the beginning of the upstroke to the incisural notch of the carotid arterial pulse; the electro-mechanical time interval (QS_2) from the onset of the QRS complex to the second heart sound; the preejection period (PEP) from the difference between QS2 and LVET. Figure 4-5 shows the expected significant increase in both systolic and diastolic pressures when the hand is placed into cold water. With the face immersion test, which is slightly more painful for most subjects, the increase in blood pressure was a bit greater than with the hand test. Heart rate and STI, however, varied quite differently with both tests. As shown in Figure 4-6, the hand immersion test caused an increase in heart rate, while the face test produced a pronounced bradycardia. If heart beat is computed on the basis of the slowest beat, we arrive at values for heart rate which are reduced to more than half the normal val-

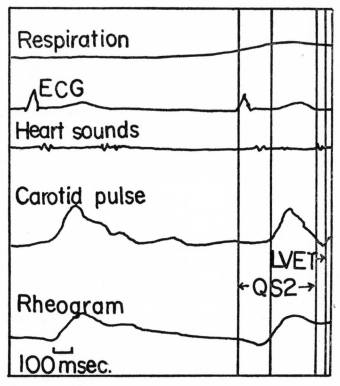

Figure 4-4. Recording of variables (ECG-lead II, heart sounds and carotid pulse) which were used for measurements of systolic time intervals (STI).

ues; Figure 4-7 illustrates the findings of a very responsive subject. This bradycardia of face immersion is due to cold stimulation and not to immersion per se, since control studies have shown that water at 30°C has no effect on heart rate. Hence, exposure of the face to cold causes an enhanced activity of both branches of the autonomic nervous system. While sympathetic stimulation results in elevation of both systolic and diastolic blood pressure, the bradycardia results from a vagal reflex through trigenimal nerve stimulation. While respiration is not affected by either hand or face immersion, the bradycardia of the face immersion test is somewhat regulated by the respiratory phase. To illustrate this the following experiments were done with cold air blown on the face. Incidently,

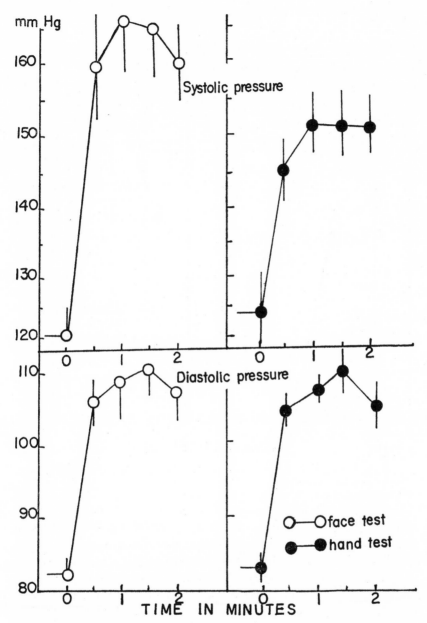

Figure 4-5. Effects of immersion of the hand or the face in water at 4°C for 2 minutes on systolic and diastolic blood pressure. For the face test the subject was breathing freely through a snorkel. Immersion of the face or of the hand in water 20° had no effect on either blood pressures or heart rate (128).

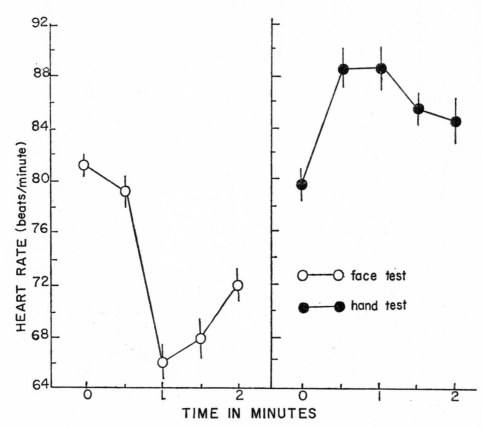

Figure 4-6. Effects of immersing the hand or the face in water at 4°C for 2 minutes on heart rate (128).

cold air was found to be as effective in producing bradycardia as cold water immersion thus indicating that the stimulation of the face by cold is the cause of this vagal reflex.

On these same subjects measurements were made of the different cardiac systolic time intervals. Stimulation of skin receptors by cold is a painful stimulus which increases sympathetic nervous system activity causing a marked elevation in blood pressure. When the hand is immersed into cold water, cardiac acceleration is observed, but marked bradycardia results from the face immersion test. Both these results suggested changes in systolic time intervals

ECG (D₂)

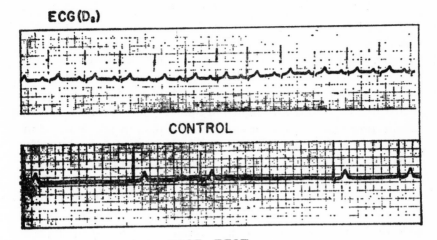

CONTROL

FACE TEST

Figure 4-7. Comparison of ECG before and during face immersion test (4°C for two minutes) (128).

with the hand or with the face immersion test. Generally, adrenergic cardiac acceleration causes a decrease in QS2 and LVET accompanied by shortening of PEP (87). Adrenaline infusion reproduces these changes but noradrenaline at a dose which causes a diminution of heart rate, seems to have opposite effects; that is a prolongation of systolic time intervals. On the other hand elimination of cholinergic effect by atropine, accentuates adrenaline actions and converts the effects of noradrenaline into a typical adrenergic responses characterized by cardiac acceleration and shortening of the STI (161). These results suggest an opposite effect of cholinergic stimulation on systolic time intervals. Further studies by Raab et al. (176) confirm this since carotid sinus pressure, which induces a vagal stimulation of the heart, causes a negative chronotropic effect and increase in QS2, LVET and PEP. All these observations indicate a correlation between changes in heart rate and variations in systolic time intervals; positive or negative chronotropic effects causing respective shortening or lengthening of systolic time intervals. Our results show that the systolic time intervals are not significantly modified by the hand immersion test but that the face immersion test causes a very significant increase in LVET and QS2 and a decrease in PEP as illustrated in Figure 4-8. Furthermore significant correlations are shown in Figures 4-9, 4-10 and 4-11, between QS2, LVET and PEP, and the heart rate. To find out whether the changes in the different systolic time intervals during the face immersion test could be explained totally by the reflex bradycardia, the equations shown in Figures 4-9, 4-10, and 4-11 were used to transform the systolic time interval data by taking into consideration the heart rate. These transformed results, corrected for heart rate, are reported in Figure 4-12. In this figure it is seen that while QS2 is no more affected by the face test, values for both LVET and PEP still remain significantly different from normal values. Consequently it may be concluded that all changes in systolic time intervals caused by autonomic nervous system stimulation cannot be explained exclusively on the basis of a chronotropic action on the heart. Similar findings were reported with increases in stroke volume due to enhanced ventricular filling (86, 218). Thus while bradycardia might explain some of the effects on systolic time intervals, other actions are involved. This complex stimulation of systolic time intervals by the face immersion is possibly related to a dual activation of both sympathetic and parasympathetic system as evidenced by increase in blood pressure and slowing of the heart respectively. In any event while the hand immersion test has no effect on systolic time intervals, the face immersion test increases LVET and decreases PEP, changes which indicate an enhanced ventricular performance.

The response to the hand immersion test conforms with the classical reactions to a stress as first described by Cannon (29), but the marked slowing of heart rate with the face immersion test is a response which may seem to have questionable physiological importance. Irving, et al. (107) have explained the bradycardia in diving seal as a defense mechanism. In cold water the intense vasoconstriction cuts off peripheral circulation, and the marked bradycardia reduces oxygen requirement while providing sufficient irrigation for heart and brain. In humans the face is also a sensing element which triggers bradycardia in the cold. However, oftentimes the face is exposed to cold while the rest of the body is kept warm. In

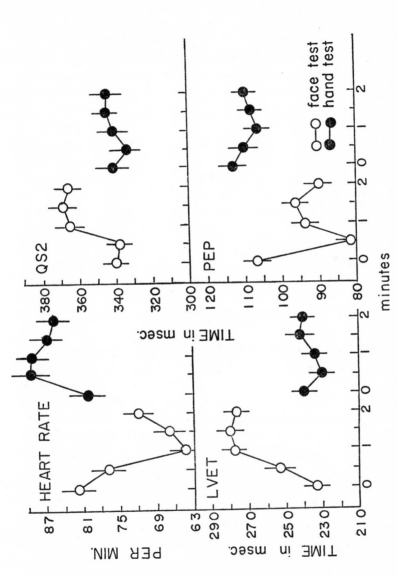

Figure 4-8. Effects of immersion of one hand or of the face into cold water (4°C for 2 minutes) on heart rate, left ventricular ejection time (LVET), electro-mechanical time interval (QS2) and preejection period (PEPE) (128).

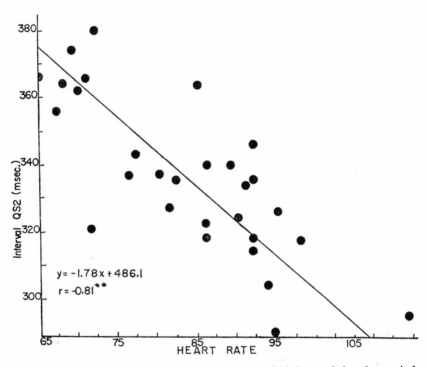

Figure 4-9. Correlation between heart rate and QS2 interval for the period preceding the immersion tests (128).

this situation the slowing of the heart seems to serve no specific purpose. The question is even raised as to whether this response may not have deleterious effects in some individuals. The pain experienced by angor pectoris patients when the face is exposed to cold winds may well be related to cardiac bradycardia; the marked increase in blood pressure at a time when heart rate is decreased imposes an additional load on the heart of these subjects and this may produce some degree of anoxia comparable to that experienced with excessive exercise.

INDIVIDUAL VARIATIONS IN RESPONSES TO FACE OR HAND COOLING

The common observation of individual differences in the response to a cold pressor test is illustrated in Figure 4-13 (222). It

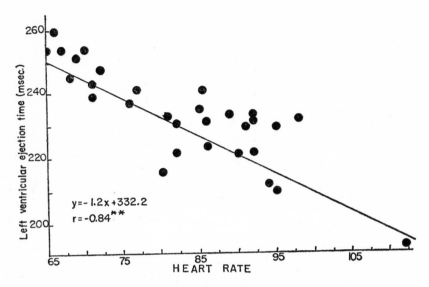

Figure 4-10. Correlation between heart rate and left ventricular ejection period (LVET) for the period preceeding the immersion tests (128).

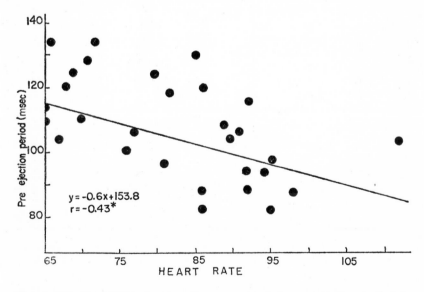

Figure 4-11. Correlation between heart rate and the preejection period for the period preceding the immersion tests (128).

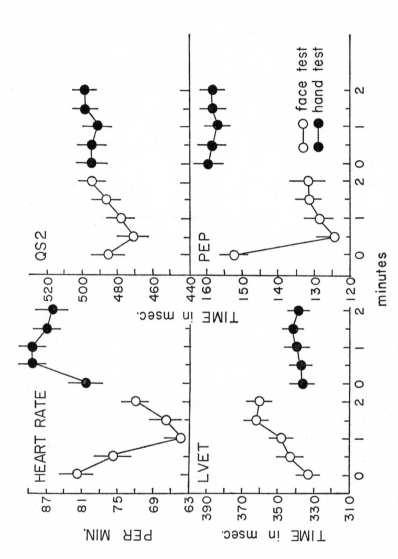

Figure 4-12. Effects of immersion of one hand or of the face in cold water (4°C for 2 minutes) on heart rate, left ventricular ejection period (LVET), electromechanical time interval (QS2) and prejection period (PEP) after these values had been transformed by referring to Figure 4-9 in order to eliminate the effect of heart rate (128).

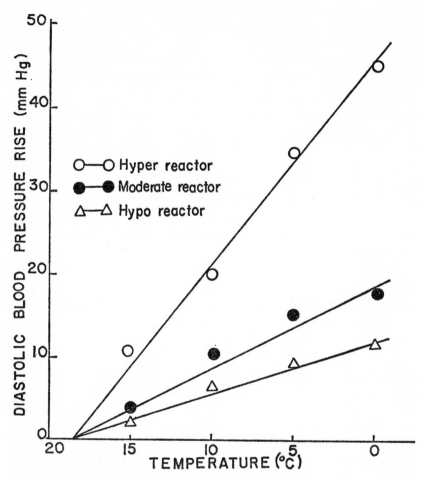

Figure 4-13. Relationship between the blood pressure elevation and the strength of cold stimulus in different individuals at various temperatures (222).

has been suggested that high response subjects have a sympathetic predominance when exposed to stress, and that in low response individuals a greater parasympathetic activity would prevail (70, 140). Individuals with excessive salivation, dry palms, slow heart rate and higher intestinal motility would have a parasympathetic dominance, and for those with dry mouth, wet palms and fast heart rate the sympathetic would be more active. In our studies since the face immersion test causes bradycardia through a vagal effect and the hand test bradycardia by adrenergic stimulation, a test of correlation was made between cardiac responses for both tests. This attempt failed to show a negative relationship between the response of these two stimulations. However, the following consideration would seem to justify another approach to this problem. Immersion of the face into cold water causes a slowing of the heart which is due to a vagal reflex action, but at the same time this test causes a sympathetic activation as evidenced by the increased blood pressures and sustained cardiac output in spite of bradycardia. One might then speculate that the effect of face immersion on the heart is the summation of vagal and adrenergic stimulation with a predominant vagal action, while the effect of hand immersion is predominantly adrenergic.

With due restriction to this type of maneuver an attempt was made at evaluating the absolute effect of vagal action on the heart during face immersion by subtracting from the measured heart rates the corresponding values obtained on each individual when the hand is immersed in water. By doing this we find, as shown in Figure 4-14, that the individual who responds most to one test, has the highest sensitivity to the other test as well. Consequently for a given subject the face immersion test would activate to the same extent both branches of the autonomic nervous system. Similarly Figure 4-15 shows that the pressor response for the hand test is very significantly related to the pressor response of the face test. These results suggest that for this type of stress, that is cutaneous cold stimulation, a high sympathetic response is accompanied by a comparable high parasympathetic response. A pertinent question would be to ask whether a high response is due to an enhanced sensitivity of the cardiovascular system or a more active autonomic nervous system. Until evidence to the contrary is reported, the latter possibility is more plausible in view of the fact that the response to cold stimulation is generally related to the degree of cold pain experienced by the subject.

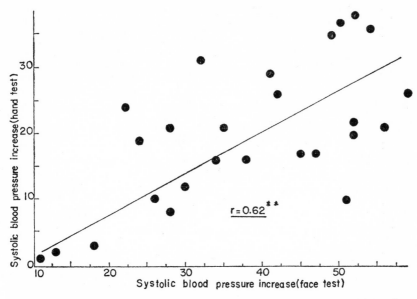

Figure 4-14. This figure deals with changes in heart rate from initial values due to either face or hand immersion test (4°C for 2 minutes). A correlation is shown between changes in heart rate with the hand immersion test and the difference between the changes in heart rate for the face test and those of the hand test. A significant correlation exists between these two variables (P < 0.05) (128).

COLD MORBIDITY

The case of sudden death upon immersion into cold water seems to be related to the present discussion. Immersion into comfortable water (30°C) while breathing through a snorkel has little physiological effect. However, breath holding in the same conditions, that is while immersed into comfortable water, causes cardiovascular responses which are completely opposite to those observed out of water. Figure 4-16 shows that breath holding in air by increasing intrathoracic pressure, diminishes venous return and as a result cardiac output. This causes a fall in blood pressure with the accompanied reflex tachycardia. However, breath holding with the

rapid heart rate 100

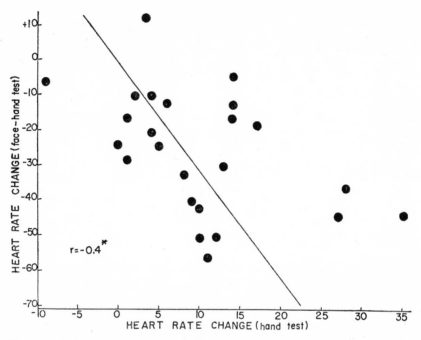

Figure 4-15. Correlation between the blood pressure increases observed with the hand immersion test (4°C for 2 minutes) and those observed with the face immersion test (4°C for 2 minutes). A highly significant correlation exists between these two variables (P < 0.01) (128).

body immersed in water at neck level increases blood pressure and causes bradycardia. It has been suggested that the hydrostatic effect of the immersion on legs and abdomen might overcome the inhibition of venous return normally produced by breath holding (85). Cardiac output and blood pressure would then increase rather than fall and this would cause reflex bradycardia. Another relevant problem also related to morbidity has to do with the effect of climate on heart attacks. De Pasquale and Burch (44) have shown a higher incidence of deaths in the summer months for southern populations, whereas winter proved more lethal in the northern part of the United States (23). Both conditions constitute a load on cardiac function. In a cold environment snow shoveling and car pushing are seemingly important factors, but the effect of cold stimulation of

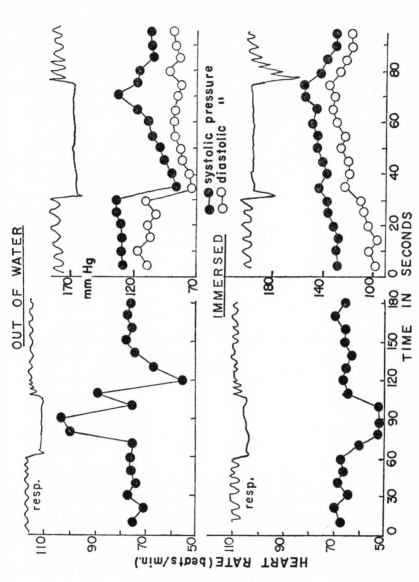

Figure 4-16. Effects of breath holding out of water and while immersed to the neck in comfortable water bath (30°C) on heart rate and blood pressure (85).

the face is possibly also involved in cardiac failure. Indeed in people
with coronary insufficiency, exposure of the face to cold could prove
fatal by imposing an excessive load on the heart. The bradycardia
of face exposure to cold by increasing the stroke volume would
cause anoxia in a heart already ischemic. In addition if exercise is
performed at the same time as the face is exposed to cold, this would
impose an additional load on the heart. However, not everything is
bad in cold climates. Repetition of stress conditions may often improve tolerance. As habitual physical activity improves heart performance, similarly moderate cold exposure of extremities and face
associated with light physical activity may indeed become a useful
adjuvant. The vigor and alertness of northern people is cited as an
example of the beneficial effect of cold weather. The evidence for
this is not altogether objective and unbiased and it may reflect primarily the opinion of these people who have adapted mentally to
colder climates. On the other hand, while flying south during the
rigor of winter may not necessarily show a kind of degenerative
adaptability, it indicates that some individuals have opted for
"flight" rather than "fight" to use Cannon terminology (29).

RESPONSES TO HEAT IN A COLD ENVIRONMENT

It would seem important while talking about cold to also discuss
some of the effects of heat. Indeed it is very common to experience
some of the effects of overheating even in a relatively cold environment, especially when a high level of physical activity is performed.
As was discussed by Burton and Edholm (25), a room at about
21°C (72°F) is comfortable for most individuals if wearing a light
suit. Deviation in air temperature by a few degrees initiates certain
responses leading to either dissipation or conservation of heat. Heat
lost by conduction or convection is directly proportional to the difference between the temperature of the skin and that of the air. For
a given sweat rate the efficiency of heat lost by evaporation is increased as the degree of humidity in the air declines. Another factor
which upsets the heat balance in a warm environment is the level of
activity. Figure 4-17 illustrates some basic cardiovascular responses
to heat (11). At 40°C and 90 percent humidity severe limitations
are imposed on heat dissipating mechanisms. At this room temper-

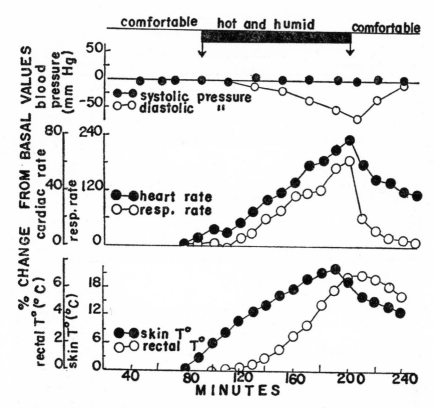

Figure 4-17. Effects of hot (40°C) and humid (90% relative humidity) on certain physiological parameters in a human subject. The cardiovascular and pulmonary responses begin to change after approximately 40 minutes of exposure to the hot and humid atmosphere at a time when rectal temperature starts to rise. The elevation of skin temperatures are much sooner and continue to increase until the end of exposure to reach levels close to the environmental temperature (11).

ature, skin temperature, being initially lower than air temperature, heat loss by conduction cannot take place. At the same time sweating becomes relatively uneffective since sweat cannot evaporate at this high humidity. The sequel of events is then a marked flushing of skin which brings a fall in diastolic pressure. This, with sustained elevation of body temperature, increases heart and respiration rate. In this environment equilibrium is not reached and dif-

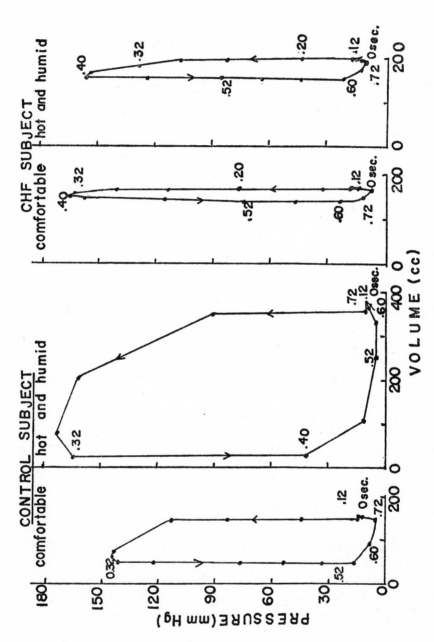

Figure 4-18. Influence of a hot and humid environment (111°F and 86% relative humidity) on volume-pressure-time "loops" of the left ventricle for a control subject and a patient with chronic congestive heart failure (22).

ferent symptoms of this stressing environment become more and more important as time progresses. A marked increase in the mechanical work of the heart results from these conditions. As seen in Figure 4-18, which represents volume-pressure-time loops, hot and humid environment brings an increase in both volume and pressure of left and right ventricles during their contraction (22). Similarly the work of the ventricle, which is a function of the pressure and volume, is markedly increased in these conditions as seen in Figure 4-19, which illustrates the load imposed on the heart in order to sustain the marked increase in peripheral blood flow caused by this stress. A more detailed analysis of cardiovascular responses to heat is given in Figure 4-20 (187). The subjects in this study were wearing suits in which water was circulated at 47.5°C. This resulted in a rapid elevation of skin temperature which was followed by a gradual increase in rectal temperature since heat cannot be dissipated fast enough and equilibrium is never attained. Vasodilatation decreases peripheral resistance and facilitates venous return and filling of the heart, thus explaining the significant increase in stroke volume at a time when both heart rate and cardiac output show marked elevation. This, of course, is very demanding on the heart as was shown in Figure 4-18. Associated with cardiovascular effects, the problem of dehydration which may develop upon longer exposure to heat is certainly an important factor of deterioration on cardiovascular functioning in a hot environment.

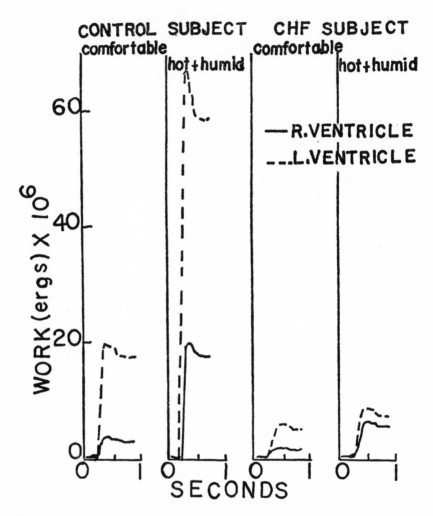

Figure 4-19. Influence of a hot and humid environment (111°F and 86% relative humidity) on the time course of accumulated work for the right and left ventricles of a control subject and a patient with chronic congestive heart failure (same subjects shown in Fig. 4-18) (22).

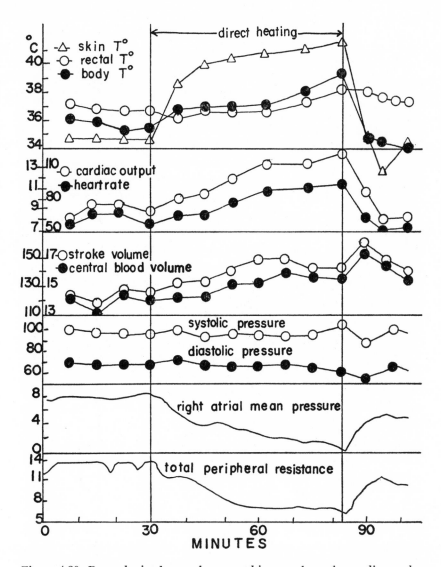

Figure 4-20. Data obtained on a human subject to show the cardiovascular responses to direct heating. The vertical scale lines divide the control period (0-30 minutes), heating period (30-83 minutes), and cooling period. Uppermost section of the graph shows the variations of temperatures. The skin temperature represents an average of leads on the thorax. Below are values for cardiac output (1/min.) and heart rate; stroke volume in ml and central blood volume in liters; systolic and diastolic blood pressure in mmHg; right atrial mean pressure in mmHg and total peripheral resistance in mmHg/1 per min. (187).

CHAPTER 5

ADAPTATION

As HUMAN BEINGS we are very generously gifted by nature. We can walk on the moon, survive an enormous degree of aggression and deprivation, wander throughout the earth in extreme conditions of weather and climates, feed on a wide variety of natural food, and live to be over seventy years of age. This is made possible by a high capacity for both resistance and adaptation to changes in the environment. Resistance may be defined as the summation of specific and nonspecific responses to a stressing situation whereas adaptation is the result of modifications in the original responses to stress. Resistance allows homeostasis; adaptation broadens the limits of resistance. There is a link of continuity between resistance and adaptation. Adaptation stimulates mechanisms which already function when the organism is first exposed to stress. If the capacity for resistance is not sufficient either because of the nature or of the intensity of the stress, a state of exhaustion is attained and disastrous reactions may lead to stress diseases or even to the destruction of the organism (192). Admittedly the limit of homeostasis borders that of pathology. The notion of specific and nonspecific reactions as applied to stress will help in defining adaptation to temperature and understanding its significance.

The specific reactions to stress are possibly the most important in the immediate resistance to stress. Indeed the effects of a cold environment are effectively opposed in homeotherms by increased heat production and decreased heat loss; resistance to hot temperature depends on increased heat loss through evaporation and decreased bodily activity; anoxia mobilizes erythrocytes; the lymphatic system is activated by infections, etc. These specific responses are the first lines of defense against the aggressor.

In 1937, Cannon described a series of responses in animals exposed to stress and was able to show that these depended on a systemic stimulation of the sympathetic nervous system (30). In 1950, Selye showed that many stressing conditions activated the secretion of hormones from the adrenal cortex (192). These early observations have stimulated many studies on neuroendocrine regulation of resistance and adaptation to stress.

I. *ADAPTATION TO COLD IN LABORATORY ANIMALS*
ADAPTATION BY CONTINUOUS EXPOSURE
TO MODERATE COLD

Much of our knowledge on cold adaptation is derived from studies done on the rat. When this animal is exposed for a few weeks at temperatures slightly above freezing, important changes are observed. The capacity for heat production is markedly increased and irreversible hypothermia is observed at temperatures much lower (−37°C) than those reported for unadapted animals (−18°C). This fact is illustrated in Figure 5-1, which shows the greater capacity for heat production and the improved tolerance to cold in cold-adapted animals. The same figure shows that the shivering curve is shifted to the left on the temperature scale in cold-adapted rats indicating the presence of nonshivering thermogenesis during adaptation to cold (48, 46, 88 and 109).

Prominent Role of the Sympathetic Nervous System in the
Development of Nonshivering Thermogenesis

Stressing conditions are well known to increase the activity of the sympathetic nervous system and this is reflected by an augmentation of adrenaline and noradrenaline in the urine. This is particularly true for continuous exposures to moderate cold which causes a marked increase in catecholamine excretion as may be seen in Figure 5-2 (152). The question was then raised as to whether this excess secretion should not make the animals more resistant to noradrenaline. Indeed Burn and Rand have postulated that the concentration and sensitivity to noradrenaline are inversely related (24). Their hypothesis is based primarily on the fact that denervation or drugs which block adrenergic activity while dimin-

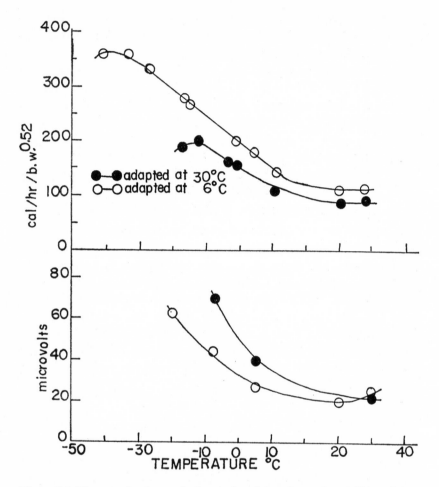

Figure 5-1. Heat production and electrical activity of muscles of 30° and 6°C adapted rats during exposure at various room temperatures (46).

ishing concentration of noradrenaline in tissues increases sensitivity to this amine. In 1957, Hsieh and Carlson reported an unexpected finding which was a turning point in the understanding of the mechanism of cold adaptation (106). They found that cold-adapted rats which secrete larger quantities of noradrenaline become more and more sensitive to the metabolic effect of noradrenaline as shown in Figure 5-3.

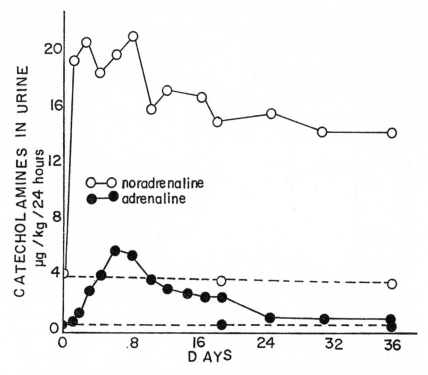

Figure 5-2. Urinary excretion of adrenaline and noradrenaline in rats (170-180 g) at + 3°C (–) and + 22°C (----). Each point represents the mean of six individual rats (152).

These results were subsequently verified and analyzed by many authors as illustrated in Figures 5-4 and 5-5 (49, 110, 111). Of course, these findings suggested that the nonshivering thermogenesis which characterized cold-adapted animals was due to this hypersensitivity to noradrenaline. Direct evidence for this came from experiments in which animals injected for a few weeks with relatively small doses of noradrenaline every day became more sensitive to this amine and at the same time were made more resistant to cold (139). Figure 5-6 summarizes these experiments. In other words, a type of cold resistance was developed in animals which were never exposed to cold but had been made more sensitive to noradrenaline by repeated injections of this amine. It is interesting

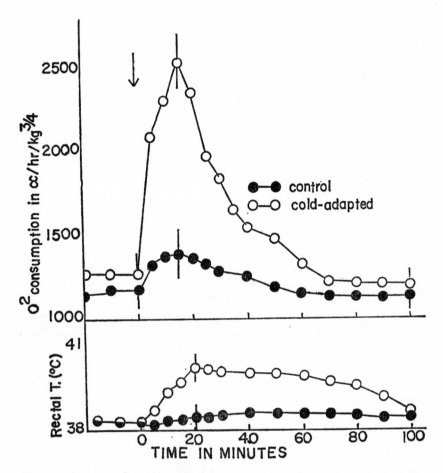

Figure 5-3. Effects of L-noradrenaline, 0.2 mg/kg, on O_2 consumption and rectal temperature of cold-adapted rats (solid circles) and warm-adapted rats (open circles). Experiments were conducted at 30°C ± 1°. L-noradrenaline was injected intramuscularly at point indicated by arrow. Vertical bars indicate standard deviation. Each point represents the mean of 4 experiments (106).

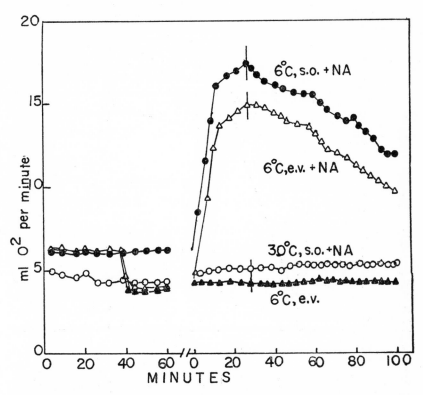

Figure 5-4. Average O_2 consumption of barbital-anesthetized cold-acclimated rats (3 per group): functionally eviscerated and infused with 0.1% sodium bisulfite in saline (6°C, EV.); functionally eviscerated and infused with noradrenaline (in 0.1% sodium bisulfite in saline) at a dose of 1 μg/min. (6°C, E.V. + NA); sham-operated and infused with noradrenaline at the same dose level (6°C, S.O. + NA). Average O_2 consumption of 2 warm-acclimated rats, sham-operated and infused with noradrenaline (30°C, S.O. + NA), is also given. Arrow indicates start of infusion which lasted 100 min. Vertical bars indicate total range of variation (49).

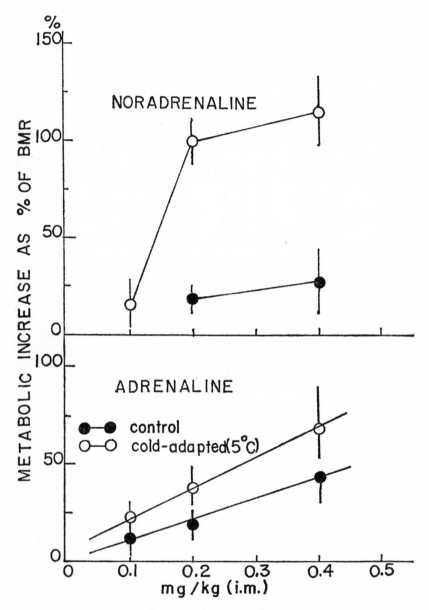

Figure 5-5. Calorigenic effect of intramuscular injections of different concentrations of noradrenaline and adrenaline in warm (25°C) and cold-adapted (5°C) anesthetized rats (111).

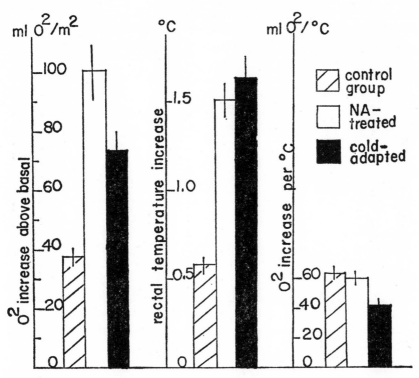

Figure 5-6. Oxygen consumption and rectal temperature increase one hour after a subcutaneous injection of noradrenaline (20 µg/100g) in control, cold-adapted rats (6°C for one month) and noradernaline-treated rats (20 µg/per day for 28 days) (139).

to note in Figure 5-7 that daily injections of isoproterenol for three weeks also produced a thermogenic sensitivity which proved as important as that seen with noradrenaline and also improved cold resistance significantly (149). Since the actions of isoproterenol are mediated primarily through stimulation of beta-receptors, these results stress their importance in cold adaptation. The mechanism of this sensitization would seem to result from the exposure of the beta-receptors to an excessive amount of the agonist. Since the action of beta-receptors is mediated through the cyclic AMP system it is postulated that this sensitization might take place at any of the following sites: at the receptor itself, the adenyl cyclase, the cyclic

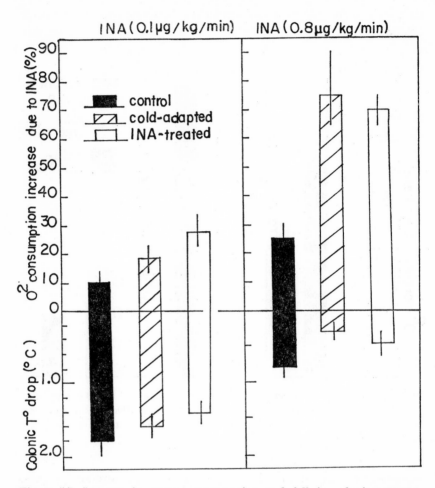

Figure 5-7. Increase in oxygen consumption and fall in colonic temperature after perfusion of isoproterenol at rates of 0.1 and 0.8 μg/Kg per
minute in control, cold adapted (6°C for 20 days) and isoproterenol-treated
rats (daily injections of 30 μg/100 g of body weight for 20 days) (149).

AMP, the kinase or any of the enzymes responsible for substrate
mobilization such as lipase, phosphorylase, etc. The activation of
some inductive enzymes is then a likely explanation. How is it possible to reconcile these results with those also showing an increased
sensitivity to catecholamines but for a completely opposite reason;

that is when the supply of noradrenaline to the receptors is reduced
(213). Closer examination of these studies indicates that in this
case it is the alpha-receptor responses that are enhanced. A pertinent question described in Figure 5-8 is to ask why excess noradrenaline sensitizes beta-receptors while the deprivation of this
substance causes a sensitization of alpha-receptors? (144). The explanation seems to be due to a difference in the site at which sensitization takes place. In the case of beta-receptors, the sensitization
would be postsynaptic, whereas for alpha-receptors the phenomenon
would take place at the presynaptic level. Accordingly with regard
to alpha-receptors, we are not dealing with a true sensitization.
What happens is that denervation or drugs inhibiting noradrenaline uptake, a most important route which deviates noradrenaline

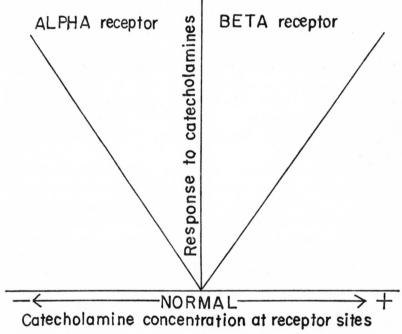

Figure 5-8. Elevated concentration of catecholamines at receptor sites by
repeated injections of catecholamines or by prolonged cold exposure, increase
the betareceptor responses, whereas a diminution of catecholamines by sympathectomy or different drugs blocking synthesis or secretion, produces an
increase in alpha receptor responses.

from receptors, potentiate the actions of noradrenaline simply by increasing the proportion of exogeneous supply to the receptor sites. The mechanisms of uptake are not changed by cold adaptation as shown in Table 5-I, and for that reason the blood pressure

TABLE 5-I

The uptake of ^{3}H-noradrenaline and ^{3}H-isoproterenol in isolated hearts of rats perfused with either ^{3}H-noradrenaline (25 ng/ml) or ^{3}H-isoproterenol (200 ng/ml). Hearts were obtained from rats exposed to cold for 3 weeks at 6°C, from rats injected daily for 3 weeks with noradrenaline (30 μg/100 g) and from control rats. (N = 10 for all groups).

Uptake (ng/g of heart)	Control	Cold adapted	Noradrenaline treated
^{3}H-noradrenaline	41 ± 3	40 ± 3	44 ± 4
^{3}H-isoproterenol	42 ± 2	41 ± 4	45 ± 3

increase in response to noradrenaline is practically comparable in normal and cold-adapted animals (129). There is also evidence that the only cardiovascular effects of noradrenaline which remain unchanged in cold-adapted animals or in animals treated chronically with noradrenaline are those mediated through alpha-receptors' stimulation. Under these conditions the response of heart rate and the force of contraction of the heart, which are mediated through beta-receptors, are significantly increased as shown in Figure 5-9 (149, 215).

These studies indicate the major importance of the sympathetic nervous system in cold-adaptation acquired by continuous exposure to moderate cold. The enhanced calorigenesis produced by prolonged noradrenaline secretion and supported by reinforced cardic functions is certainly the important characteristic for this type of adaptation. Admittedly, however, noradrenaline is not the only factor in cold-adaptation. Repeated injections of noradrenaline or isoproterenol, while producing as much sensitization to catecholamines as cold exposure, do not elevate the threshold of tolerance as much as the actual exposure of the animal to a cold environment (146). Another hormone which might be important is obviously thyroxine because of its paramount importance in calorigenesis. As was discussed previously, the fact that the thyroid is more active during cold exposure, that thyroidectomy interferes with cold-

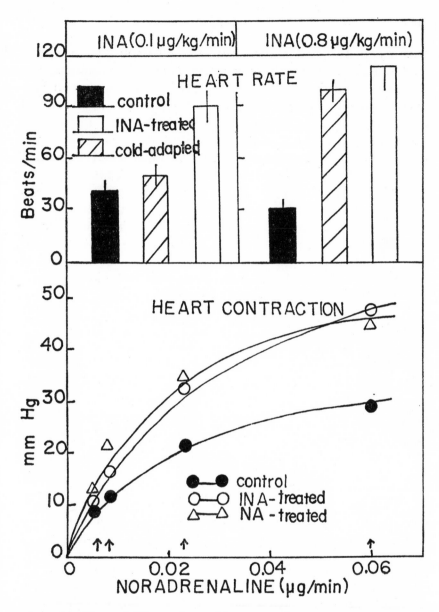

Figure 5-9. Effects of different doses of isoproterenol on heart rate of anesthetized control, cold-adapted (6°C for 21 days) and isoproterenol-treated (20 μg/day for 21 days) rats. Similar groups were used to measure the effect of various doses of noradrenaline on ventricular contraction of isolated hearts (149, 215).

adaptation, and that chronic treatment of normal animals with relatively small daily doses of thyroxine (10 uμ/day) improves cold tolerance, point to the possible importance of this hormone in cold-adaptation. If we consider that chronic injection of thyroxine and noradrenaline simultaneously, as shown in Figure 5-10, improve tolerance to cold almost as much as the actual exposure to cold, it would then seem that the combined effect of thyroxine and nor-adrenaline may well contribute to the most important fraction of nonshivering thermogenis in cold-adapted animals (146, 148). There are, however, unexplained aspects of cold adaptation. For instance, it is shown in Figure 5-10 that the cold-adapted animals are less sensitive to noradrenaline than any of the other treated groups and yet they are the most resistant. Another unexplained finding is the remarkable tolerance to frostbite in cold-adapted animals. After three hours at $-25°C$ all groups, including the thyrox-ine-noradrenaline-treated group which maintained a high colonic temperature, had frostbite of tail and extremities with the excep-tion of the cold-acclimated group in which the animals had a warm and flexible tail even six hours after continued exposure (146). This observation points to some vascular adaptation which has not been sufficiently studied. Adaptation to cold is a complex phenome-non which develops through the harmonious interdependence of various reactions interrelated within a time-dependent scheme. When all the functions of the organism are rearranged to cope with the new requirements, we are dealing with a new system, which is not necessarily different basically, but improved.

This may explain why certain secretions which were greatly stimulated when an animal is first exposed to cold are becoming less and less important as adaptation is modeled. The secretion of catecholamines, for examples of thyroxine or of corticosterone, di-minish with time of exposure while the animal simultaneously be-comes more and more resistant to cold. It is a sort of integrated systemic potentiation of all the heat producing and heat preserving systems which exist in the organism. Important changes in the mechanism of hormonal regulation of thermogenesis are described during cold adaptation. But does the substrate for calorigenesis adapt, are there changes in the rate of energy mobilization, or are there shifts in the pathways of heat production?

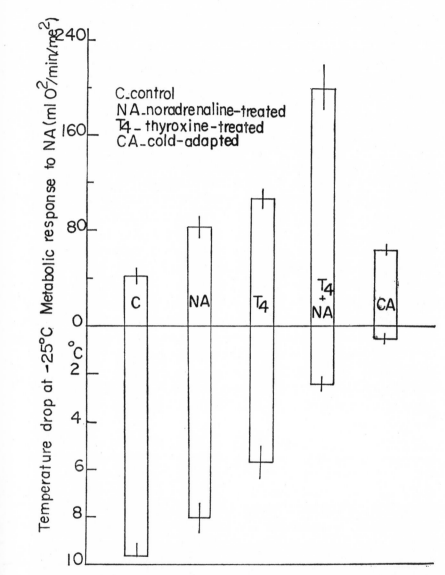

Figure 5-10. Oxygen consumption increase one hour after a subcutaneous injection of noradrenaline (30 μg/100 g) to rats treated for 35 days with noradrenaline (30 μg/100 g/day), with thyroxine (10 μg/Kg/day), with both thyroxine and noradrenaline and to rats adapted to cold (6°C for 35 days). The colonic temperature drop after 3 hours at −25°C is also reported for the various above mentioned groups (146).

ADAPTATION BY INTERMITTENT EXPOSURE
TO SEVERE COLD

Continuous exposure to moderate cold has been used with success to study hormonal interactions (65, 125). However, for wild animals, domestic animals and humans this type of cold exposure is not frequent. It is only by accident that under natural conditions an organism is exposed to a cold environment that would more than double caloric requirements and produce uninterrupted shivering for periods of hours. What is much more frequent is exposure of short duration to various cold temperatures accompanied by wind. These short exposures, which are not long enough to activate metabolic responses, cause a temporary stimulation of the sympathetic nervous system. The resulting reactions, primarily those affecting the cardiovascular system seem excessive and the question may even be asked as to whether they are not sometimes harmful. Claude Bernard was of the opinion that pain oftentimes serves no purpose (171). The pain that results from acute exposure to intense cold also seems to cause such responses as vasoconstriction, increase in blood pressure and heart rate, which are not needed for the purpose of resisting the aggression. Fortunately as for many stresses repetition brings adaptation. But there is evidence that adaptation by repeated short exposures to intense cold is different from that already described for continuous exposure to moderate cold.

Glasser and co-workers (72, 73, 74) followed the responses of the heart rate to repeated immersion of the tail of rats into water at 4°C. Their results, reported in Figure 5-11, show a gradual decline in the response to this stimulus. They also reported that rats with bilateral frontal cortical lesions failed to adapt to this situation and exhibited a continuous elevated response to cooling. These findings were given as evidence of a central control for this type of adaptation. The characteristics of this adaptation are somewhat identical to those of habituation which is described as a weaning in response intensity resulting from repeated stimulation.

While accepting the importance of the central nervous system in the development of habituation, the prominent role of the autonomic nervous system as a necessary intermediate must also be ac-

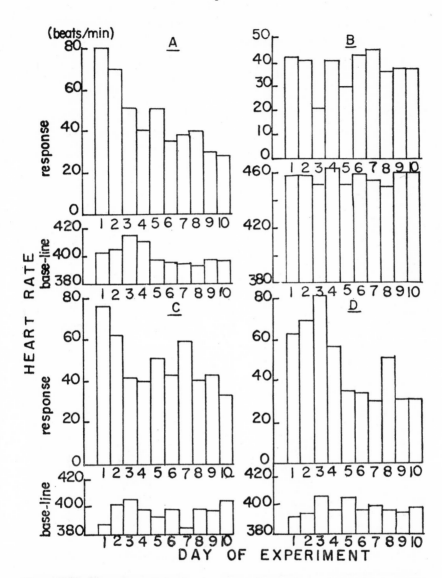

Figure 5-11. Responses of the heart rate to immersions of the tails at 4°C. A — mean of ten normal control rats; B — mean of nine rats with bilateral frontal cortical lesions; C — mean of six rats with occipito-parietal lesions; D — mean of five rats with unilateral frontal lesions (74).

cepted. In continuous exposure to moderate cold, the autonomic nervous system reinforces some defense mechanisms and thus plays a key role in the development of adaptation, whereas the diminution of its activity parallels the acquisition of adaptation induced by repeated exposure to severe cold. The evidence for this may be summarized in the following manner. First of all, Figure 5-12 shows

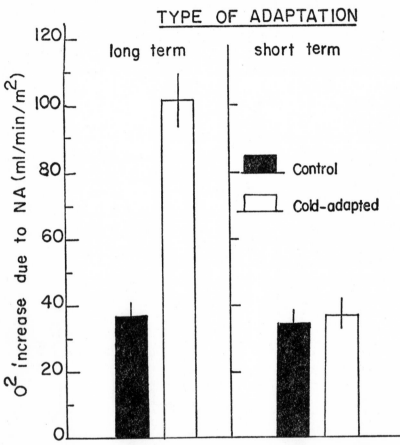

Figure 5-12. Effect of adaptation to cold on the metabolic response to noradrenaline. The oxygen consumption response was measured after subcutaneous injections of 0.5 mg/kg noradrenaline to four groups of mice. One group had been exposed for 10 minutes at −10°C every hour during the day for a total of 9 exposures per day for 3 days (short-term adaptation); a second group was exposed at 10°C continuously for 6 weeks (long-term adaptation) and the other two groups were used as respective controls (143).

the substantial difference in sensitivity to noradrenaline between
the two types of adaptation: in adaptation resulting from repeated
exposure to severe cold, enhanced sensitivity to noradrenaline is not
a major factor (143). This is probably because the noradrenaline
secreted, as shown in Figure 5-13, is relatively small compared to
that reported with adaptation resulting from continuous exposure
to moderate cold (147). Repeated exposure to intense cold reduce
the stressing responses and minimize the strain on the body. To
estimate the importance of habituation, it is essential to find out
how the habituated animal responds to prolonged exposure to cold.
Experiments on rats reported in Figure 5-14 show that habituation
improves tolerance to cold (142). Some habituated animals do not
develop irreversible hypothermia even seven hours after exposure
to −20°C, and as seen on the same figure, these same animals have
survived this test more than twenty-four hours. It is interesting to
note—and this has been confirmed many times—that only a certain
percentage of the animals become habituated to cold; similar findings have been reported with regard to the acquisition of reflex
conditioning. However, when dealing with metabolic adaptation, it
is very exceptional to find an animal which fails to show improvement to severe cold. It should also be mentioned that metabolic
adaptation produces a greater cold resistance than habituation as
seen in Figure 5-15; on the other hand habituated animals were
only exposed for a total of three hours at −20°C, whereas the other
adapted animals had been in the cold for two months at 6°C (129).
In mice, as shown in Figure 5-16 and 5-17, it was possible to quanti-
tate quite accurately the degree of habituation: The lower the temperature at which habituation was acquired, the better the survival
in the cold at −5, −10 or −15°C. Habituation did not protect the
mice at −20°C because at this extreme temperature the maximum
average survival time is about one hour (143).

As mentioned previously noradrenaline does not seem to be
very important in habituation, although the excretion of this
amine is slightly higher in the habituated animals both during the
period of adaptation and during the cold test (Fig. 5-13) (147).
Measurements of urinary adrenaline gave quite different results.
While the control animals showed an eightfold increase in adrenaline during the four hours exposure at −16°C, the excretion in

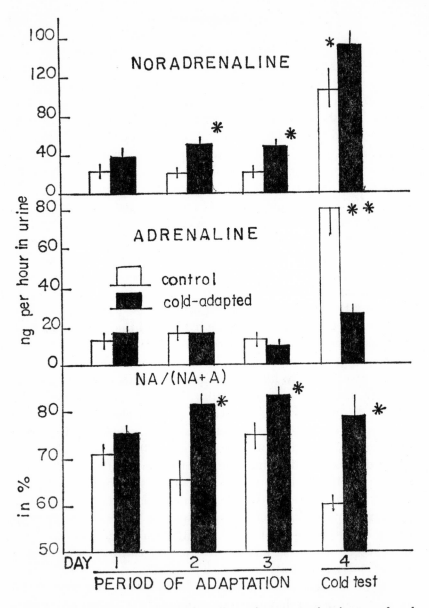

Figure 5-13. Adrenaline and noradrenaline urinary excretion in control and cold-adapted rats (10 minutes at −20°C at hourly intervals for a total of 9 times per day during 3 consecutive days) during the adaptation period and during the four hour exposure at −16°C on the fourth day. Means ± standard error are given; one asterisk indicates statistical significance at the 5% level and the double asterisk at the 1% level (147).

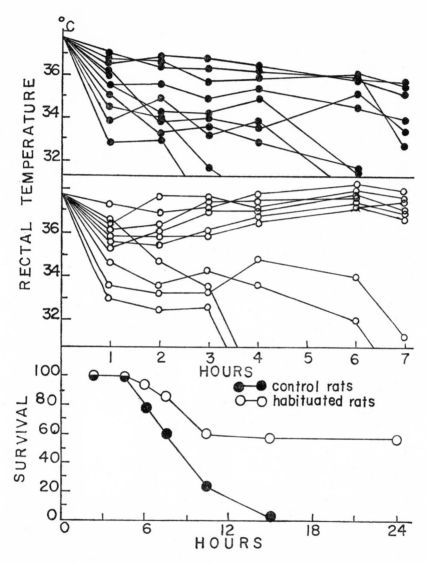

Figure 5-14. Changes in rectal temperatures and survival time at −20°C of control rats and of rats adapted to cold by hourly exposures of 10 minutes at −20°C for a total of 10 times per day over a period of 10 days 142).

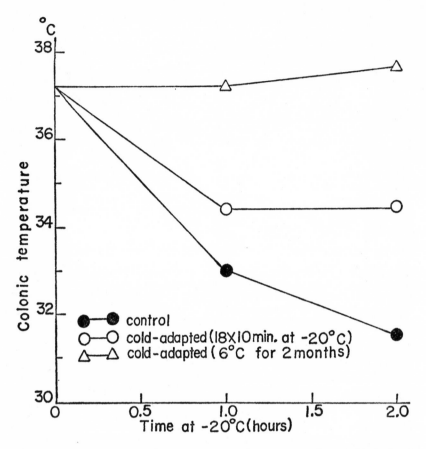

Figure 5-15. Fall in colonic temperature of rats exposed at −20°C for 2 hours. Three groups of rats are compared: one control, one adapted by exposures of 10 minutes at −20°C at hourly intervals for a total of 9 exposures per day during 3 consecutive days, and one group adapted by continuous exposure at 6°C for 2 months.

habituated rats remained almost normal. The increase in adrenaline points to a direct involvement of the adrenal medulla. The gland is known to be activated by stimulation of different parts of the central nervous system. Some investigators have shown that a variety of stresses selectively activate adrenaline or noradrenaline secretion in humans (58, 57, 59). Von Euler in a recent review con-

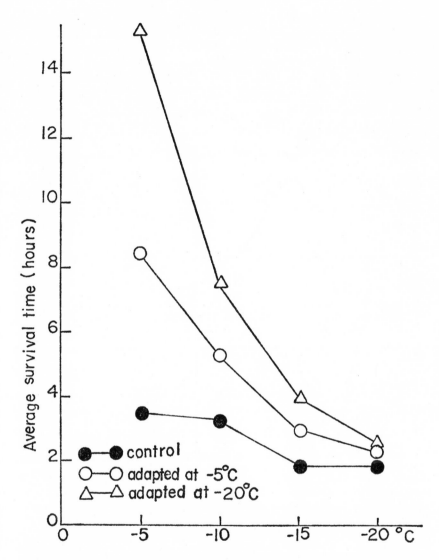

Figure 5-16. Effect of adaptation at different temperatures on survival time. Mice were adapted at −20, −5 and +25°C by being exposed for 10 minutes at these various temperatures at hourly intervals for a total of 9 exposures per day during 2 consecutive days. Survival time was measured for these 3 groups at temperatures ranging from −5 to −20°C (143).

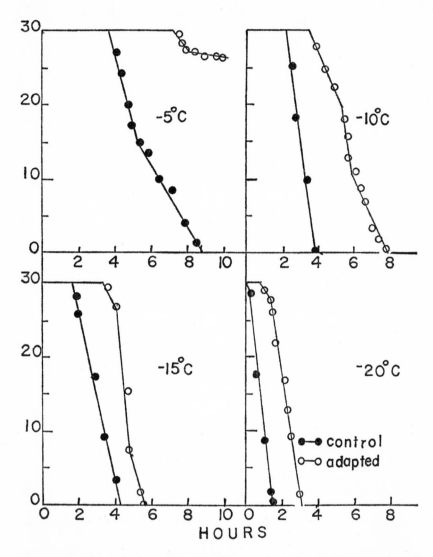

Figure 5-17. Effect of adaptation to cold on survival time. Mice were exposed for 10 minutes at −5, −10, −15 or −20°C at hourly intervals for a total of 9 exposures per day during 2 consecutive days. On the third day survival, at the temperature at which mice had been adapted, was measured and is compared to that of control mice (143).

cluded that emotional stresses of all kinds stimulate the secretion of
the adrenal medulla (60), but there also is evidence that this action
is reduced by habituation. Indeed it has been shown that adrenaline
excretion, which becomes very elevated in humans exposed for the
first time to a centrifugation test, drops to within normal range
after the sixth run (66). In parachute jumping, however, this type
of habituation does not occur and the adrenaline excretion remains as high in trained subjects as it is in those making a first
jump (59). From these observations von Euler (59, 60) concludes
that habituation does not take place if an element of real danger
persists, thus suggesting that some emotional components play a
major role in this type of adaptation. The results reported in Figure 5-13 would then indicate that repeated short exposures to
severe cold eliminate the emotional part of this stress. The habituated animal becomes more resistant to prolonged severe exposure.
However, there is no evidence that this is due to an enhanced
thermogenesis resulting from the development of new or improved
factors. Simply what may be happening is that initially the response
to cold is excessive and deleterious to the animal; after repetition,
the emotional component diminishes and the overall energy is used
sparingly to overcome the physical effect of cold. In this conception
the secretion of adrenaline in nonhabituated animals would be excessive and deleterious. This interpretation, while speculative, finds
support in the observation where restraining animals in the cold,
that is adding emotional to physical stress, greatly aggravates the
severity of the stress. The other possibility is that repeated exposures to severe cold activate some metabolic processes sufficiently
to improve tolerance to cold. Although the sensitivity to noradrenaline is unchanged, the secretion of this amine is increased in
adapted animals (Fig. 5-13) and the levels of plasma thyroxine are
changed following adaptation. This could suggest the importance
of a combined calorigenic action of noradrenaline and thyroxine to
explain the delayed onset of hypothermia in adapted animals as is
shown in Figure 5-18. Assuming comparable heat loss, Figure 5-19
shows that an increased heat production of approximately 15 percent is sufficient to delay hypothermia in adapted animals. Repeated
injections of thyroxine over a period of three days improved cold

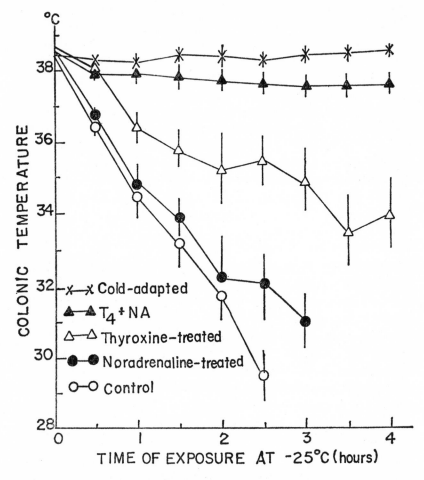

Figure 5-18. Colonic temperature variations of various groups of rats when exposed at −25°C for a period of 4 hours. Standard errors of the means are indicated (146).

tolerance, and if one compares these results with those reported in Figure 5-18, it may be seen that approximately the same number of injections, whether they are given over a period of three days or three weeks, gives a somewhat comparable degree of enhanced tolerance to cold. Thus these findings suggest that the role of thyroxine is one of primary importance whatever the type of adaptation.

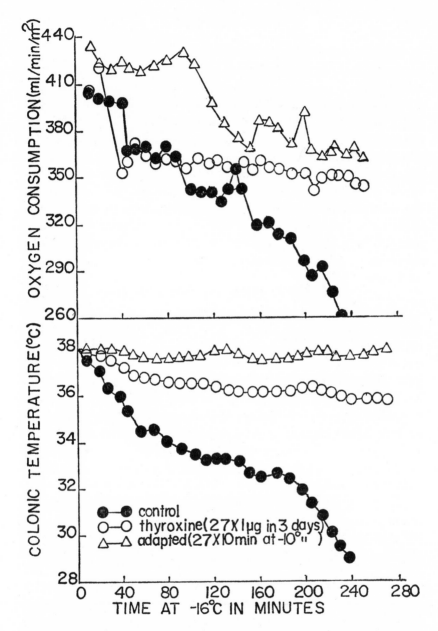

Figure 5-19. Effect of exposure at −16°C for 4.5 hours on oxygen consumption (STP) and colonic temperature (measured by telemetry) of control, thyroxine-treated and cold-adapted rats. Each point represents the mean value of 8 animals (147).

Whether adaptation comes from continuous exposure to moderate cold or from intermittent exposure to severe cold, it is always characterized by an enhanced metabolic activity. Furthermore repetition of severe cold decreases the emotional component of this stress. If survival in the cold is considered a valid criterion for adaptation, then improved calorigenic capacity, is of course, important whereas the role of attenuated emotional responses has not been fully estimated. On the other hand, if the criterion for adaptation is not survival in the cold but attenuation of cold pain, reduction of disturbance enhanced efficiency to perform or decreases cardiovascular responses so habituation more than metabolic adaptation becomes of primary importance. This will be the basis of discussion of the analysis of studies done on humans.

II. *ADAPTATION TO COLD IN MAN*

Nude man is probably the least fitted animal to live in a cold environment. Without protection, humans would have been most comfortable only by migrating annually between temperate and tropical zones. Technology has expanded the frontiers of human populations to the point that cities are now being built in the Arctic where until recently a few thousand Eskimos had managed to survive through ingenious developments of clothing and shelter. Human beings cannot solely be considered physiological entities; it is important to determine and evaluate the technological, physiological and psychological components of adaptation of humans to a cold environment.

In recent years many experiments have been performed on different groups or populations and it would seem justified to analyze the evidence for adaptation on the basis of length and severity of cold exposure as was done for laboratory animals.

ADAPTATION BY CONTINUOUS EXPOSURE
TO MODERATE COLD
Primitive People

In recent years extensive studies have been made on groups that have lived nude or poorly clothed in a cold environment for many centuries. The *aborigines of Central Australia* are the most typical of these populations and were studied in the thirties by Hicks and co-workers (76, 97) and some twenty-five years later by Hammel, Scholander, and others (79, 195). They sleep naked at temperatures of 0°C or below in the winter, without shelter and between small fires. Their responses to overnight exposure at 3°C were compared to those of white subjects and are reported in Figure 5-20. In these conditions the white subjects could not sleep and started shivering very rapidly; within one hour the oxygen consumption had increased significantly and the total heat production kept cycling above basal levels for the duration of the eight-hour period. Concommitant with their extra heat production, the rectal temperature fell during the first hour but remained stable at slightly above 36°C for the rest of the night. The average body temperature also dropped rapidly for the first four hours but thereafter remained fairly stable at about 30°C. These subjects displayed typical responses to moderate cold, i.e. enhanced heat production through shivering, decreased heat loss through lowering of skin temperature and a slight fall in rectal temperature. A steady state was achieved at the latter part of the night when the different components of the heat balance were kept constant. Completely different responses were observed in the aborigines. They failed to respond to cold. They did not shiver and their skin and rectal temperatures continued falling throughout the night. They never reached a steady state, and yet what is most important, they could sleep all night. Like hibernators on the long winter night, the aborigines oppose cold in the most economical way while remaining able to recuperate.

Another group which has been studied recently is the *Bushmen of the Kalahari Desert* (81, 224). Like the aborigines they wear little or no clothing and are exposed at night to temperatures nearing zero in July. When exposed to overnight cold as seen in Figure 5-21 they behave like the aborigines: they do not shiver and their body temperature keeps falling for the duration of the exposure. Their skin temperature does not seem to fall as rapidly as was noticed in the aborigines but this may be explained on the basis of adipose tissue insulation which is much less important in Bushmen. So this is another group which resists cold, not by producing more

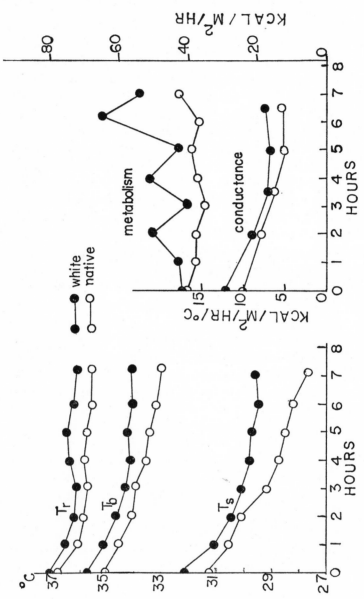

Figure 5-20. Average thermal and metabolic responses of six Central Australian aborigines and four control white subjects during night of moderate cold exposure in the winter season. Tr, rectal temperature; Tb, mean body temperature; Ts, average skin temperature. The air temperature was about 3°C. Effective insulation of blanket bag and surrounding air in use by native group = 3.4 clo. (redrawn from 98).

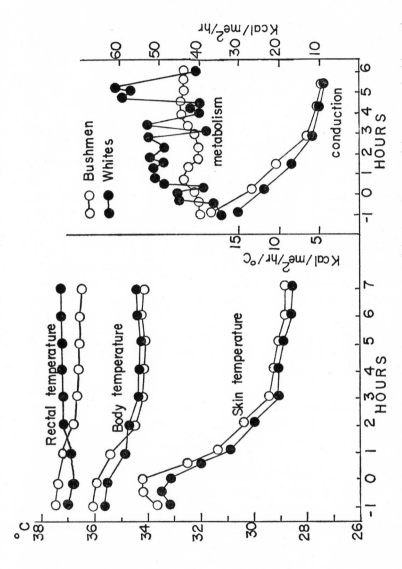

Figure 5-21. Average thermal and metabolic responses of 14 Kalahari Bushmen and four control white subjects during a night of moderate cold exposure in the winter season. Air temperature 3 to 8°C. Effective insulation around native group = 2.4 clo. (81)

heat, but by cooling down slightly to a level which is compatible with sleep and yet never endangering the organism.

The Alacaluf Indians of Tierra del Fuego is a third group which has been studied with regard to its response to cold exposure (80). Like the other primitive populations mentioned, the Alacaluf Indians do not increase their heat production when exposed to cold overnight. However, since their metabolic rate is already high at the beginning of the exposure, the degree of hypothermia is less pronounced than in the Aborigines or the Bushmen.

If we consider the case of *the Ama,* the women divers of Korean Peninsula, we see that they also adapt to cold. The temperature of the water in which they dive falls to 10°C in the winter, and yet they continue their daily work throughout the year (102, 180). In the course of a dive the oral temperature falls by 2 to 4°C and they differ essentially from nondivers by showing a considerable elevated shivering threshold as seen in Figure 5-22. This population behaves like the primitive groups discussed thus far: shivering is reduced and there is a fall in body temperature; in a sense one can say that the thermostat is lowered to a more economical level.

Most interesting observations were made on the Ainu, the Japanese women divers. Itoh compared the calorigenic action of noradrenaline (0.025 mg/10 kg body weight subcutaneously) in Ainu and Japanese students (108). Figure 5-23 shows a significant increase of oxygen consumption, plasma free fatty acid and plasma ketone bodies in the Ainu group fifteen and thirty minutes after the noradrenaline injection; no difference is noted in the Japanese student group. From these studies a negative correlation was found between the energy metabolism and the plasma FA level in the Ainu group, while this correlation was positive in the Japanese students. This suggested to the author a faster oxydation of FFA with higher elevation of the metabolism in the Ainu group. On the other hand, a positive correlation was found between energy metabolism and plasma ketone bodies in response to noradrenaline. For these reasons the authors assume that fats are more utilized in the form of ketone bodies than in that of FFA itself in the Ainu. These studies are possibly the only cases of enhanced response to noradrenaline in humans adapted to cold. Therefore it can be said

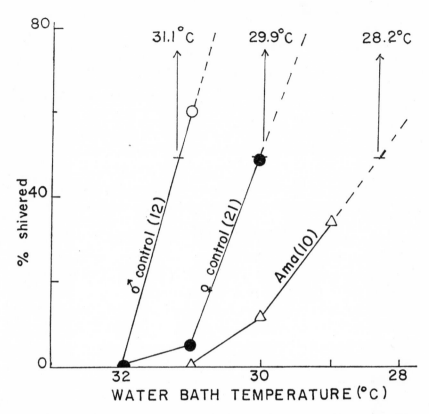

Figure 5-22. Incidence of shivering in Ama and control subjects at various water temperatures. Numerical figures in parenthesis indicate number of subjects (102).

that when sufficiently exposed to cold, man behaves like other species and can adapt to cold through some metabolic changes. These groups are exposed to cold for long periods of time throughout the year and for most of their lifetime. What happens if white subjects are similarly exposed for prolonged periods to moderate cold?

White Populations

In 1954 we were able to test this question by following the responses to cold of a group of ten soldiers who were transferred

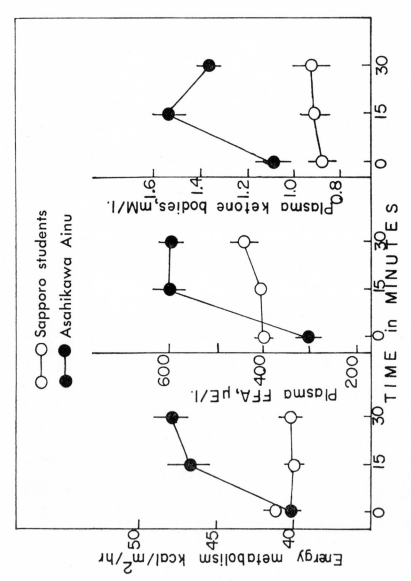

Figure 5-23. Effects of noradrenaline on energy metabolism, plasma FFA concentration and plasma total ketone bodies Asahikawa Ainu and Sapporo students (108).

from Winnipeg to Northern Manitoba in November and remained there until the next spring (133). From January to April the average air temperature was −20°F. These soldiers lived outdoors from eight to ten hours per day, six days per week, throughout the winter. In November, January and March they were exposed naked for one hour at 47°F. In March a control group, composed of subjects exposed at the most one half hour per day to outside temperature, were subjected to the same test. Figure 5-24 shows a decline in shivering and internal body temperature throughout the winter, while the fall in skin temperature remained comparable for the same period. Heat production is decreased and this is reflected by a larger drop in body temperature, a response quite similar to that observed in Aborigines. Another group of white subjects was studied by Davis et al. (42). Figure 5-25 shows the responses of white subjects to a standard cold exposure in October to those measured in February. They reported a decrease in shivering, although in their study the changes in rectal temperature were not significantly different throughout the winter.

All these studies on primitive populations or on white subjects show similar patterns for cold adaptation. Prolonged exposure to cold reduces shivering; this seems to be the essential feature of cold adaptation. This reduction in heat production means a saving of approximately 40 percent on energy stores. This could not really be an important saving for the organism; more energy is dissipated simply by walking. So this response of the organism is very likely explained on the basis of other priorities. If life is not endangered, the responses of the body seem to be oriented towards retaining individual entity, preserving homeostasis, avoiding unnecessary challenges or, if we want to express it this way, cherishing self-comfort: the pursuit of utopia is within our flesh. The evidence on hand suggests that adaptation to cold in humans might be due to habituation. The first response to cold is to increase heat production. Even if a subject is told and explained that the cold exposure will only last one hour and that at 15°C or so, even naked, it is not necessary to shiver, because the body temperature would only drop a few degrees centigrade, which does not endanger the organism, just the same this subject does shiver. The body has control of the situation.

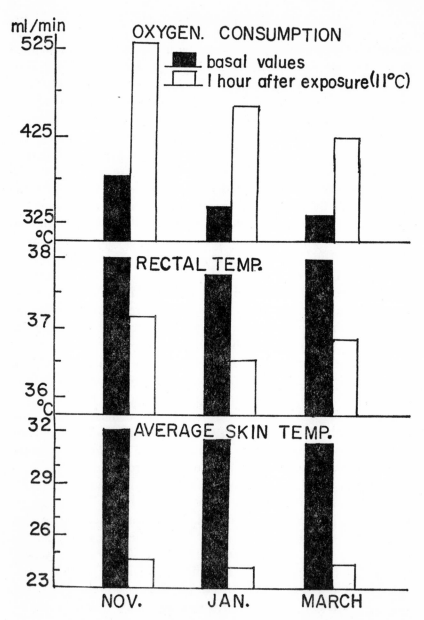

Figure 5-24. Variations in oxygen consumption, rectal temperature and average skin temperature of eight soldiers exposed for one hour at 49°F in the nude in November, January and March. These soldiers lived outdoors all winter at Fort Churchill, Manitoba. Canada (average air temperature for January, February and March was −25°F) (133).

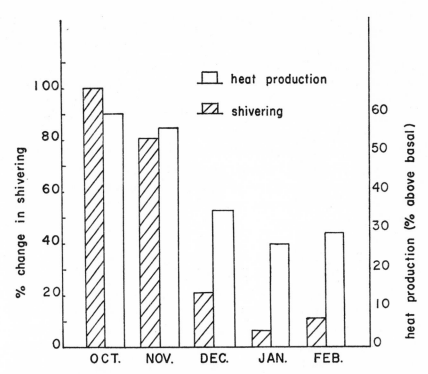

Figure 5-25. Shivering and heat production response to a standard cold exposure (one hour) of 6 subjects as a result of seasonal climatic change. Cold exposures were made once monthly. The mean monthly outdoor temperature varied linearly between +12 and −3°C during that period (42).

This is fine because the mind can be fooled. It is only after repeated cold exposure, which delimits the nature of the aggression, that the mind learns through the body to recognize that the overall response of the organism is excessive; that the shivering response, which is very disturbing and uncomfortable, is not needed. It is then evident that adaptation to cold in humans is very closely related to that observed in laboratory animals. Shivering is the first line of defense but nobody likes it. In laboratory animals exposed to rather severe prolonged cold, it is replaced by nonshivering thermogenesis mediated by noradrenaline. In humans exposed to relatively less severe cold, shivering also disappears and this causes some degree of hypothermia. However, in man there is no evidence of substitution by nonshivering thermogenesis. The differences between species do not really exist however.

The rat at 6°C, the temperature at which most studies on adaptation have been done, has to increase heat production by over 200 percent. Consequently blocked this response would lead to pronounced hypothermia becoming rapidly irreversible. Consequently shivering has to be replaced by some other forms of heat production which prevents too great a fall in body temperature. In humans the increase in heat production is only 40 percent so that shivering can be cut off and not replaced by other mechanisms for heat production; the hypothermia that results does not become deleterious in any way.

ADAPTATION BY INTERMITTENT EXPOSURE TO SEVERE COLD

There are groups who in their natural habitat are also exposed to cold but in a manner which is rather different from that described so far. We have seen that some populations live in the cold, nude or with very little clothing. When exposed to a standard cold test, they shiver much less and become slightly more hypothermic than control unadapted subjects. There are groups of people which also live and work in very cold environments with the difference being that they usually wear sufficient clothing to cope with the situation.

Eskimo Populations

The first example that comes to mind is that of the Eskimo population that lives in the Northern territories. All studies concur to show that the Eskimos, when exposed to a standard cold test, respond exactly the same as non-Eskimo subjects; indeed they begin to shiver at the same skin temperature (2). When compared to the Kalahari Bushmen (98), it was found that the Eskimos shiver more, that they keep a higher skin temperature and maintain a higher rectal temperature in the cold. In 1963, Rennie (180) reviewed the evidence on the subject and could not find any difference between thermoregulation in Eskimos and white people. "They begin to shiver at the same skin temperature. They begin to perspire at the same skin temperature, and their overall tissue insulation during maximal vasoconstriction, though significantly less than in the control subjects of this study, is what would be predicted for nonacclimatized American males of comparable fat thickness" (180).

Indeed Figure 5-26 shows that for a given insulation, tissue conductance is the same between Eskimos and non-Eskimos; conse-

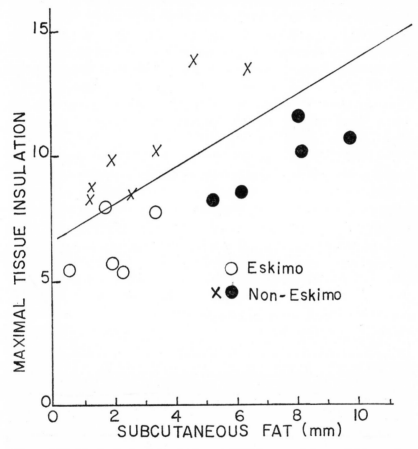

Figure 5-26. Maximal tissue insulation of 5 Eskimos and 5 non-Eskimos plotted as a function of mean subcutaneous fat thickness. These subjects were studied in the summer of 1960 in Barrow, Alaska. The remaining 7 non-Eskimos (x) were University of Buffalo students studied in the summer of 1961 (180).

quently, the reduced total insulation observed in the Eskimos is not
due to some vascular adaptation but simply to reduce subcutaneous
fat in the Eskimos (180). Overall the Eskimos respond to cold
much more like unadapted subjects. In a sense this is not surprising. Contrary to the Australian Aborigines and other primitive
groups, the Eskimos have succeeded in protecting themselves so
effectively that they are probably seldom exposed to temperatures
conducive to shivering or prolonged cold discomfort. The unique
design of their clothing has been adopted in recent years by white
populations. After some time spent in the Arctic, it rapidly becomes evident that shivering and total body cooling are not often
experienced if proper clothing is worn. However, what is much
more frequent is the stimulation of the face and of the extremities.
Different types of face masks and goggles have been tested in the
Arctic, but they have never been very popular and are not worn
generally. Similarly while the rest of the body might be comfortable
it is always difficult to protect the feet adequately, especially if a
person is not engaged in physical activity. For these reasons, as concluded previously, the Eskimos show no evidence of adaptation
when subjected to an overall exposure to cold. Cold exposure of the
extremities, however, gives quite different responses.

Over twenty-five years ago a Canadian expedition in the Arctic
was organized by Brown and co-workers from Queen University
(19). Among the many tests and measurements that were made,
skin temperature and blood flow of the hand were determined during immersion in waterbaths at temperatures between 5 and 45°C.
Figure 5-27 shows that for all temperatures tested blood flow is
significantly higher in the Eskimos than in white subjects; this difference is sustained since it was persistent over the two-hour exposure. This response resulted in improved performance and reduced cold sensation in the Eskimos. The authors of these studies
on Eskimos observed "their ability to work with bare hands and to
continue to perform fine movements in the winter cold" and re-
marked that "at 10°C waterbath, the Eskimo experienced a sensation of coldness upon immersion but there was no pain, and the
three subjects studied were asleep within twenty minutes. The four
subjects in the control group experienced first a sensation of severe

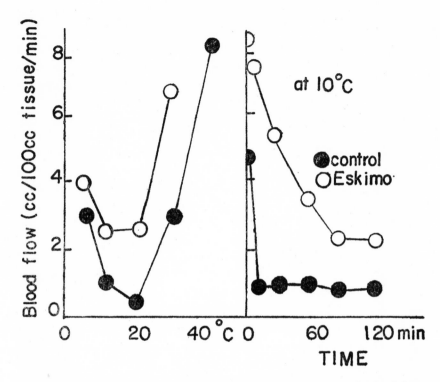

Figure 5-27. The left part of the figure shows the average hand blood flow at various water-bath temperatures for control white subjects and Eskimos. Each point is the average of the readings made every five minutes during the last hour of a two-hour exposure period. The right part of the figure shows the effect of immersion of hand in water-bath at 10°C (19).

coldness in the immersed arm, and a deep, aching pain developed which reached maximum intensity in about three minutes. None of the control group was able to sleep." These observations have been confirmed and new evidence has been obtained recently by Eagan who emphasized this difference by using cold air (−22°C) instead of cold baths (56). As seen in Figure 5-28, all Eskimos used in his study sustained quite easily the thirty minutes of finger cooling and retained a skin temperature above 20°C for the greater part of the experiment. Only 25 percent of the white subjects could finish the experiment; the other subjects suffered frostbite or could not tol-

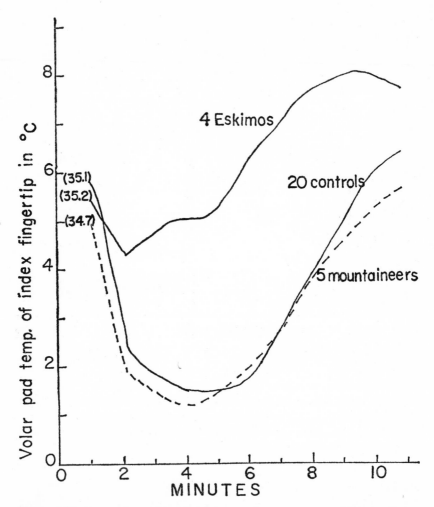

Figure 5-28. Comparison of group averages of temperatures of fingers during immersion in water at 0°C under standard test conditions. Pre-immersion finger temperatures are shown in brackets. Fingers were immersed at 0 minute (56).

erate the most intense cold pain. All evidence confirms the remarkable tolerance of the Eskimos to cold exposure of the extremities. A likely explanation is that Eskimos develop this special adaptation by repeated exposure of the extremities to severe cold.

This being the case we would expect that white subjects, if exposed to comparable conditions, would also become more tolerant. This has been the subject of many studies, but before reviewing these it may be pertinent to analyse further the situation with regard to the Eskimo.

Concommitant with the enhanced circulation and reduced cold sensation, immersion of the hand into cold water brings a reduced activation of the sympathetic nervous system in the Eskimos. Figures 5-29 and 5-30 show that systolic blood pressure of Eskimos is not very much affected by a cold immersion test. These results were obtained on adult subjects, children and Eskimo women of Igloolik in the Northen territories. This is a small village of approximately 800 Eskimos where many of the Canadian IBP (International Bi-

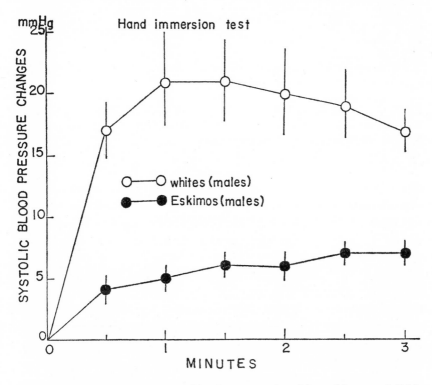

Figure 5-29. Changes in systolic blood pressure in white subjects and Eskimos during immersion of the hand in water at 4°C for 3 minutes (150).

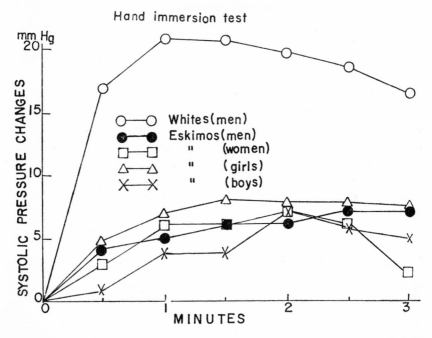

Figure 5-30. Changes in systolic blood pressure in white male subjects and in male, female, young boy and girl Eskimos during immersion of the hand in water at 4°C for 3 minutes (150).

ology Program) studies have been done (150). The population of this community is in a period of cultural transition in the sense that the Eskimos live in heated little houses and are adopting some of the eating habits of the white people. As a result, the women and the children who attend heated comfortable schools are certainly not exposed to cold as much as they have been used to—probably not much more than comparable white populations living in subarctic or cold temperature regions. Consequently, it is a bit surprising to realize, as shown in Figure 5-30, that this part of the Eskimo population is almost as tolerant to cold as the adult hunters. These findings raise the question as to whether the adaptation in Eskimos is aquired or is the result of some inherited ethnic characteristics.

White Populations

On the other hand, there is definite evidence of adaptation in white populations.

Gaspé Fishermen

Studies have also been carried out on a group of subjects who are exposed to special cold environmental conditions (135, 136, 137). These are the fishermen of the Gaspé peninsula who fish in the area of the Gulf of Saint-Laurent between April and December when the average water temperature for that period is 9.4°C and that of the air 12.4°C; during the fishing season they immerse their bare hands into cold water for many hours while pulling nets and unhooking fish. Their responses to a cold immersion test were compared to those of a control group composed of manual workers. Both systolic and diastolic blood pressure increase were significantly smaller in the fishermen, as shown in Figure 5-31, than in the control subjects. Figure 5-32 shows that the skin temperature of the immersed hand remained higher in the fishermen and this is due to a greater heat flow. Immersion of the hand into cold water was an unpleasant and painful stimulus for the control subjects, so much so that three of them had a fainting reaction; the fishermen, however, retained their characteristic joyful mood throughout the procedure. These fishermen are comparable to Eskimos in the sense that they show minimum response to exposure of the extremities to cold. It is also interesting to note that both groups maintain a better heat flow in the hands immersed into cold water; this response, of course, may well have significance in explaining their cold tolerance. The Gaspé fishermen, like the Eskimos, gave no evidence of adaptation when exposed naked at 15°C for one hour. On the contrary they shivered more than the control subjects as seen in Figure 5-33. The results reported in Chapter 1, showing that upon exposure to cold the intensity of shivering for a group of subjects is greater in individuals with higher skin temperatures, suggests that the Gaspé fishermen shiver more in the cold because they retain a higher average skin temperature. Like the Eskimos, the fishermen maintain a higher skin heat flow in the cold but this has a mixed effect

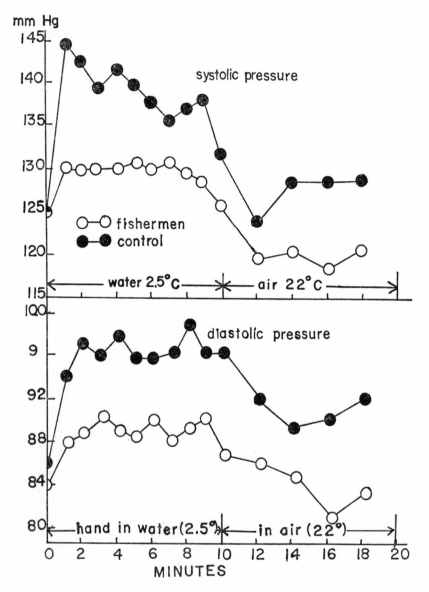

Figure 5-31. Variations in systolic and diastolic blood pressure of fishermen from Gaspé in Québec and control subjects during immersion of one hand in cold water (135).

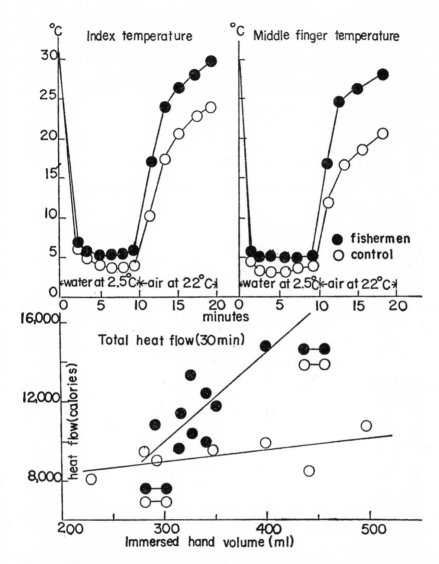

Figure 5-32. Variations in skin temperatures of index and middle finger and correlation between immersed hand volume and heat flow in Gaspé fishermen and control subjects during immersion of the hands in a water bath at 5°C (136).

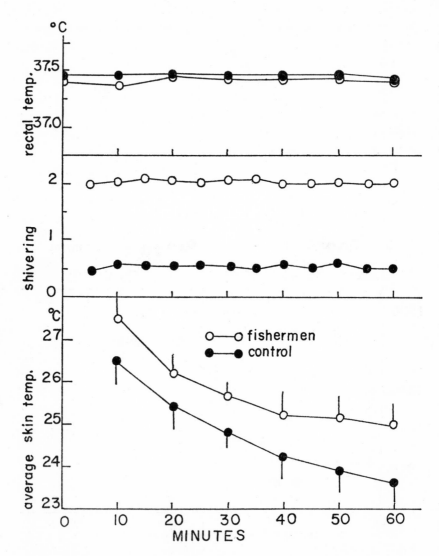

Figure 5-33. Rectal temperature, average skin temperature and shivering intensity of Gaspé fishermen and control subjects exposed nude for one hour in a room at 15°C (137).

on their response. When the extremities are placed in the cold, the pain sensation is largely alleviated and they show little response, but if the whole body is exposed to cold, the opposite response is observed as they shiver more than the control subjects.

This illustrates once more the specificity of adaptation. A given stimulus causes specific conditions of disturbance accompanied by discomfort or pain. The object of adaptation is to oppose the stressing situation which is repeatedly experienced. The adaptation of Eskimos, aborigines or fishermen is different because the nature of the aggression, the condition in which cold is experienced, is different from one group to the other. The organism is accordingly conditioned to specifically defined situations. Furthermore, all experiments reported show the negative nature of this conditioning. In populations that experience prolonged overall exposure to cold, the absence of shivering is typical, whereas in Eskimos and fishermen the essential characteristic is the remarkable tolerance to acute cold. Since these adaptive responses tend to eliminate the disturbing effects of cold, that is shivering and pain sensations, it would seem as though the initial responses to cold were exaggerated. This reasoning is true only in retrospect. Indeed until adequate knowledge of the nature of the stress condition is learned by the individual exposed to cold, shivering and cold pain provide useful warnings against an aggression which can always be deleterious or even fatal. The initial reactions constitute the essence of the alarm reaction of Selye (193) or the orientation reaction of Pavlov (174). While repeated exposures to cold in humans result in the disappearance of the initial responses, in other species adapted to much more severe conditions, these reactions are supplemented or replaced by enhanced cellular reactions such as an increased sensitivity to noradrenaline observed in rats (106) and dogs (206).

Many successful attempts have been made to reproduce evidence of cold adaptation in human subjects (41, 42, 104, 133, 206). In 1958, Glaser, et al. observed a greatly diminished response to the hand immersion test (4°C) following repeated exposures (73). No evidence of any localized change in the hand was noted, and the authors concluded that the underlying mechanism appeared to be habituation that is a reversible process which depends on the mind

and which involves the diminution of normal responses or sensations. Following experiments on the rat to study the effect of
frontal lobe ablation on the process of habituation, Griffin (78) reported some studies in lobectomized human subjects. As may be
seen in Figure 5-34, repeated immersion of the hand in water at 4°C
abolished the rise in blood pressure in normal and schizophrenic
subjects but had no effect on the response of lobectomized subjects;
lobectomy prevented adaptation. It was concluded from these

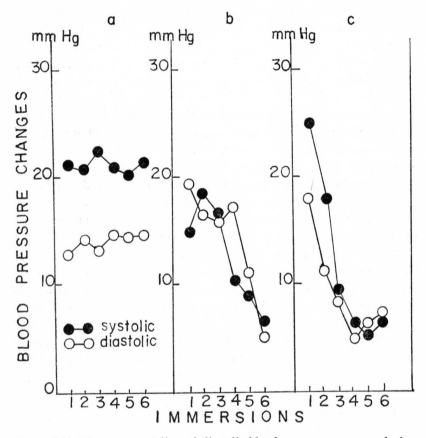

Figure 5-34. The mean systolic and diastolic blood pressure responses during
six successive periods of localized cooling on the initial day of the experiment in (a) 7 leucotomized subjects, (b) 3 schizophrenic control subjects and
(c) 6 normal control subjects (78).

studies that the frontal areas of the cerebral cortex have a role in the development of habituation.

Student Population

Other approaches have been described in an effort to demonstrate the participation of the central nervous system in adaptation to cold. An experiment was designed in which the test subjects were repeatedly exposed to the following situations (141). In the first group one hand was immersed into cold water (4°C) for 2.5 minutes, twice a day for nineteen days. As may be seen in Figure 5-35, the test became less and less stressing with repetition and this is evidenced by the gradual decline in blood pressure responses. The second group of subjects was exposed to the same conditions as the first group but in addition they were given a mental arithmetic test simultaneously. In this group Figure 5-36 shows that the scoring improved with time and that the response to both tests (cold and mental arithmetic tests) also decreased significantly. The interesting point is then to see whether this diminished response is due to an enhanced tolerance to cold, a greater facility to do the arithmetic test or a specific adaptation to both tests simultaneously. The answer to these questions is given in Figure 5-37. At the end of the adaptation period, the subjects who had been exposed to both tests throughout the experiment were exposed to these tests separately. When comparing the responses to either the cold test or the arithmetic test, before and after the adaptation period, no significant differences could be seen. In other words the subjects adapted very specifically to the situation to which they were exposed. If exposed to cold only, they became more tolerant to cold; if exposed to cold and the mental arithmetic test they became adapted to these tests simultaneously, but when tested with cold alone they gave no evidence of adaptation. Considering the results on performance for the arithmetic test and the blood pressure responses, the results of these experiments suggest the following interpretation. When the experiment was initiated, the mental arithmetic test was not sufficient at first to distract the subjects from the cold-water test. Consequently, the blood pressure was high and scoring low. But after repetition, more attention was given to the mental test and the sub-

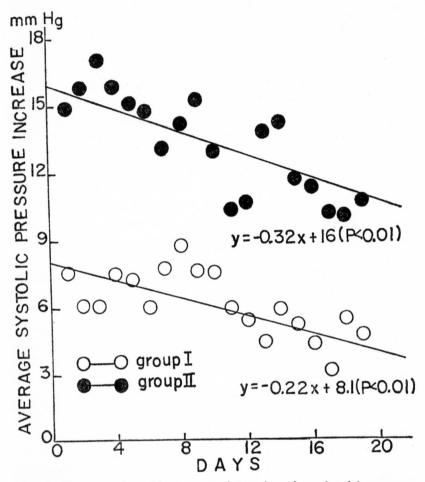

Figure 5-35. Effect of a cold water test (immersion of one hand in water at 4°C for 2.5 minutes) on systolic blood pressure increase above basal values in two groups of subjects. Group I was exposed to the cold water test twice daily for 19 days and group II was exposed to the same procedure as the one used for group one but in addition the subjects of this group were subjected simultaneously to a mental arithmetic test. Each group was composed of 8 subjects (142).

jects were distracted from the cold test. At that point blood pressure decreased and the scoring improved. It seems possible to say that in one group (those exposed repeatedly to the cold-water test alone)

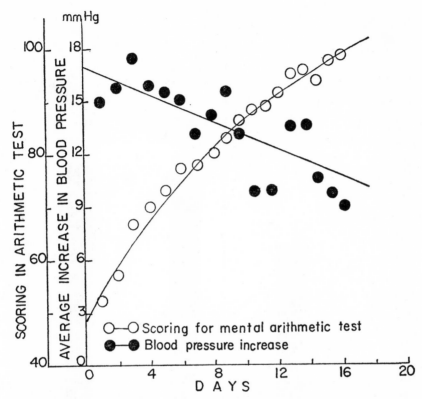

Figure 5-36. Daily maximum increases in systolic blood pressure and average score in a mental arithmetic test in group II which was exposed to a cold water test (4°C for 2.5 minutes) and to a mental arithmetic test simultaneously twice a day for 19 consecutive days (141).

habituation to cold has developed, whereas in the other group (those exposed to the cold and arithmetic tests simultaneously), adaptation to cold per se has not developed; instead increased attention to the mental test has distracted the subjects from cold. In both conditions the subjects have become more tolerant to cold but different mechanisms appear to be involved. In the same experiment some results were obtained which confirm even further the concept of specificity of adaptation to cold and illustrate the intricacy of the relationship between the environment and its mental and sensory perception. For all tests described thus far the left hand

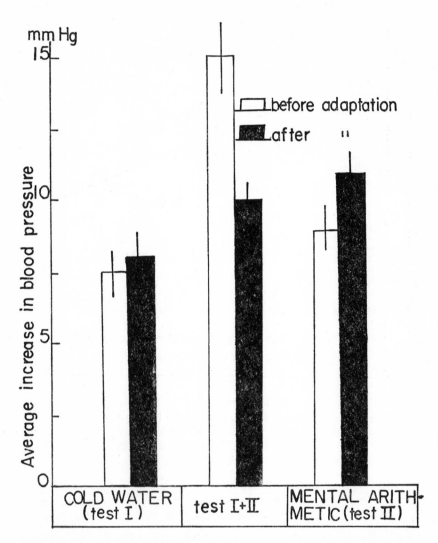

Figure 5-37. Average increase in systolic blood pressure when the mental arithmetic test and the cold water test were given before and at the end of the adaptation period in the group which had been adapted to both tests simultaneously. The figure shows that this group adapted to both tests simultaneously but not to either test when given separately (141).

was used. At the end of the experiment and one month later, the response of the right hand, exposed to cold water for the first time, was measured and compared to that of the adapted left hand. Figures 5-38 and 5-39 show in both experimental groups mentioned previously, a much greater response in the right nonadapted hand, even greater than that recorded for the left hand at the beginning

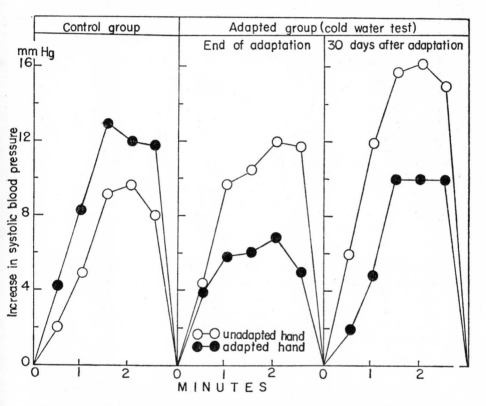

Figure 5-38. Increases in systolic blood pressure during a cold-water test in a group of subjects adapted to the mental arithmetic test and to the cold water test simultaneously. The measurements were made at the end of the adaptation period and one month later. The responses of the left adapted hand is compared to that of the right nonadapted hand. A highly significant difference (P<0.01) was observed between both hands and this was due to a very high response in the unadapted hand which could not be predicted. Indeed in control group (unadapted) both hands gave identical responses (141).

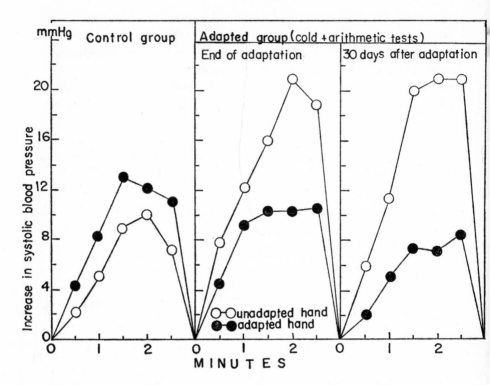

Figure 5-39. Responses of systolic blood pressure to a cold-water test in the group adapted to the cold-water test. The measurements were made at the end of the adaptation period and one month later. The responses of the left adapted hand is compared to that of the right nonadapted hand. As was shown in figure 5-37., with the other group, a highly significant difference (P < 0.01) was observed between both hands (141).

of the experiment. Subjectively the pain sensation experienced was also much greater when the right hand was immersed as compared with the left. The same figures show that even one month after the end of the experiment the response to the immersion of the unadapted hand remains surprisingly greater than could be expected.

There is certain analogy between these results and those reported for Gaspé fishermen. In both cases the specificity of adaptation was observed. But when adaptation is acquired for a given set of conditions, changing the procedure completely alters the response. The fishermen who are adapted to immerse their hands into cold water, shiver more than control subjects when exposed naked at 16°C for one hour, and the subjects adapted to the immersion of the left hand into cold water show an unexpected increased response when the right hand is tested. The fact that this reinforced reaction is still present after one month indicates the central origin of this phenomenon. These results have also shown that adaptation is very specific and that it is characterized by a diminution of the original responses to the aggression; in other words, adaptation to cold is acquired through habituation.

CHAPTER 6

INTERSTRESS ADAPTATION

THERE IS A CERTAIN ELEMENT of specificity in cold adaptation in the sense that the degree of tolerance is somewhat related to the type of exposure that has been experienced. We can talk conveniently about three levels at which adaptation can develop: the physiological, the psychological and the behavioral. Physiological adaptation resulting in an enhanced heat production does not seem to be important in humans. As was discussed in Chapter 5, it is not because the mechanisms for this to occur are not existing, but rather because exposure to moderate cold is not sufficiently long in duration to activate the responses leading to enhanced calorigenesis. Another type of physiological adaptation seems to develop quite readily in both humans and laboratory animals and it is controlled by some psychological mechanisms. Repeated exposures to severe cold, which are often experienced in the Northern hemisphere, lead to a psychophysiological adaptation which has been called habituation. This permits the organism to withstand the cold while alleviating the sensation of cold discomfort. Another way of resisting cold, which is also important, is to learn to live in the cold, to become the apprentice of nature. The way to work, or to eat, or to derss are all important things to learn when exposed to any type of environment. Learning to wear two pairs of mitts instead of one because the extremities are most susceptible to freezing, but at the same time, learning to wear one sweater instead of two to avoid, in conditions of increased muscular activity, excessive accumulation of sweat into the clothing.

An example of behavioral adaptation is illustrated in the way the Eskimos have learned to protect themselves from the cold. The double layer clothing provides sufficient insulation for exposure to extreme conditions and yet the possibilities of ventilating their garment permits work in less severe conditions without excessive sweating. Adaptation is then a mixture of behavioral and psycho-physiological responses which make it possible to live comfortably in a cold environment. The processes involved in this adaptation point to rather specific mechanisms. The point is then raised as to whether adaptation to a given environment or a specific stressing situation, will improve or deteriorate resistance to other physical or psychological stresses.

COLD AND ALTITUDE

Colder temperatures and high winds are more frequent in high mountains than at sea level. For this reason a certain number of studies on cross-adaptation have compared exposure to altitude and to low temperatures. Another reason is that these stresses lend themselves quite well to quantitative estimation. Adaptation to altitude by repeated exposure has been reported by many investigators using a variety of tests. Fregly has used the righting reflex to determine the degree of adaptation (68). This test measures essentially the length of time an animal remains conscious when exposed to high altitude. It has been shown that daily exposures to 39,000 feet for two weeks increases the righting time by a factor of 3 as seen in Figure 6-1. It is interesting to note that adaptation produced by this method did not cause any significant changes in either hemoglobin concentrations or hematocrit values. In a sense this is not unexpected since the daily exposures to altitude were of very short duration (between 1 to 5 minutes). These results indicate that repeated short exposure to high altitude result in a type of adaptation which can be compared to that observed with cold exposure. Adaptation to both stresses is obtained by very short exposure to extreme conditions (142, 145). This is illustrated in Figure 6-2 which shows increased tolerance to cold ($-20°C$) and to altitude (30,000 feet) by repeated short exposures to these stresses. We have already indicated in Chapter 5, that this type of adaptation has no effect on noradrenaline sensitivity in contrast with the enhanced response to noradrenaline in animals adapted by continuous exposure to moderate cold (CM adaptation to cold). On the

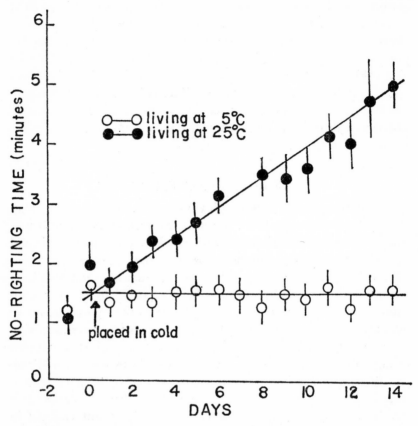

Figure 6-1. Effect of daily exposures to 39,000 feet simulated altitude on the no-righting time of 16 rats living in air of 25°C and 16 others living in air of 5°C. Standard error is shown for each mean (68).

other hand, intermittent exposure to severe altitude produces adaptation without the increase in red blood cells and hemoglobin concentration which normally results from continuous exposure to moderate altitude (CM adaptation to altitude). The conclusion to be drawn from these studies is that while adaptation to cold or altitude is possible by continuous exposure to moderate conditions (CM adaptation), adaptation by intermittent exposure to more severe conditions (IS adaptation) is also possible but through different mechanisms. The most specific responses to

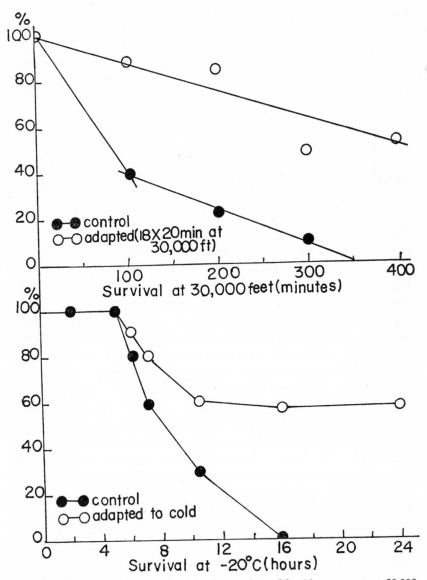

Figure 6-2. Survival time of a group of rats adapted by 18 exposures at 30,000 feet simulated altitude over a period of 2 days. Each time the animals were brought up to this altitude over a period of 8 minutes, they were kept there for a period of 4 minutes and brought down to sea level in 4 minutes. In the lower part of the figure is represented the survival time at −20°C of rats adapted to cold by hourly exposures of 10 minutes at −20°C for a total of 15 exposures over a period of 2 days. Controls were also used for each type of adaptation (142, 145).

CM adaptation to cold (increased sensitivity to noradrenaline) and to CM adaptation to altitude (increased hemoglobin concentration) are not observed with IS adaptation to cold or to altitude. It would seem as if the specific changes produced by CM adaptation are not important in IS adaptation. These observations suggested studies to test the possibilities of interstress adaptation to cold or altitude in animals adapted by either CM or IS type of exposures.

Considering first the effect of cold adaptation on tolerance to altitude, it has been shown that the type of adaptation is important. Animals adapted to cold by intermittent exposure survive longer at 30,000 feet, but those exposed to 6°C for two months show no improved tolerance to altitude as shown in Figure 6-3 (145). Fregly has even shown that animals kept at 5°C cannot adapt to altitude (68). Consequently, prolonged exposure to moderate cold fails to improve resistance to cold and in some conditions an impaired tolerance has been reported. On the other hand, as seen in Figure 6-3, repeated exposures to severe cold (6 daily exposures of 10 minutes at −20°C over a period of 3 days) markedly increase tolerance to altitude (145). In fact the enhanced resistance to high altitude is as pronounced in cold-adapted animals as it is in animals adapted to altitude; the percentage of animals surviving more than 400 minutes at 30,000 is comparable in altitude- adapted animals (Fig. 6-2) to that reported in Figure 6-3 for cold-adapted animals. This, of course, indicates the implication of nonspecific factors in adaptation to altitude. The mechanisms involved in this type of adaptation are not known. One possibility is that repeated exposure to either severe cold or altitude eliminates the emotional component of the stress. Indirect evidence for this was obtained by showing a decreased adrenaline secretion after repeated exposure to severe cold (147). As an explanation for the enhanced tolerance to altitude produced by either cold or altitude exposure, it is tempting to suggest that a decrease in the emotional component of the stress, with the resulting diminution of sympathetic stimulation, might play an important role.

Consequently repeated exposures to severe cold improve tolerance to altitude while continuous exposure to moderate cold has the opposite effect. A possible explanation for this latter observation may be related to some metabolic action. The oxidizing activity of

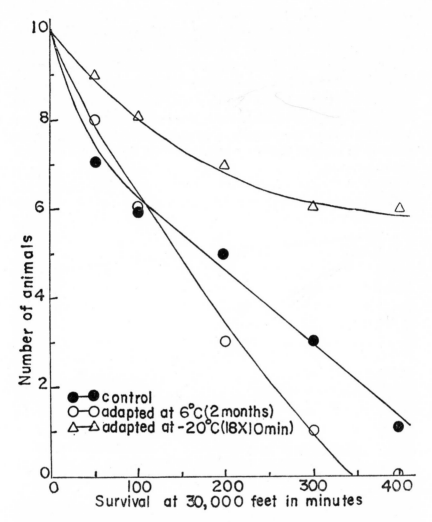

Figure 6-3. Survival of 3 groups of rats at 30,000 feet. One group was control, one had been adapted to cold (6°C for 2 months), and another had been adapted to cold (18 periods of 10 minutes at −20°C) (145).

cold-adapted animal is enhanced as evidenced by the increased metabolic rate, enhanced thyroid activity and the greater sensitivity to catecholamines as was shown in Chapters 2 and 5. Of course, at altitude the limiting factor is the availability of oxygen to the tissues. Consequently, if an anmial is hypermetabolic, and this seems to be the case in cold adaptation, the available oxygen will be used up more rapidly with the consequence that deleterious effects will appear sooner. Some results reported in Figure 6-4 show that exposure at −20°C for two hours completely cancels the effects of previously acquired tolerance to altitude by repeated exposure to severe cold (145). This finding may also be explained by the hypermetabolic state of the animals or by some other factor such as exhaustion of adrenal steroids, suggested by studies of Selye or other unknown causes.

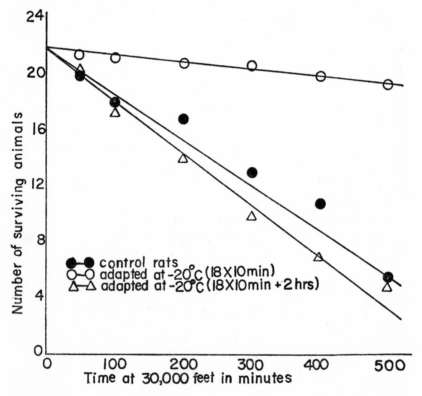

Figure 6-4. Survival of 3 groups of rats at 30,000 feet. One group served as control, one had been adapted to cold (18 periods of 10 minutes at −20°C) and another had also been adapted to cold in a manner similar to that of the second group but in addition was exposed at −20°C for 2 hours the day before the survival test at altitude (145).

Studies dealing with the effects of adaptation to altitude on cold tolerance have shown opposite results. Indeed Figure 6-5 shows no improvement of cold tolerance in animals adapted to altitude by repeated exposure to 30,000 feet. There is no explanation for these results which stress the complexity of the mechanisms involved in both specific and nonspecific adaptation to stress.

As discussed in Chapter 4, the cardiovascular effects of the cold-pressor test are due to an activation of the autonomic nervous system. It is interesting to note that adaptation to altitude or even acute hypoxia (12,500 feet simulated altitude for one hour) brings a significant depression of the cold-pressor response (169). The following factors are suggested as possible explanations for these changes: a decreased P_{CO_2} at the vasomotor center due to hypoxic hyperventilation, diminished P_{O_2} of medullary centers or alterations of the threshold of peripheral receptors.

COLD AND EXERCISE

The effect of physical fitness on cold resistance has been studied specifically. Untrained subjects were submitted to a physical training program for three weeks. Using successive minute pulse recovery rates after exercises on a treadmill, the scoring shown in Figure 6-6 indicates the effectiveness of the program (3). The responses of these subjects to a standard cold test (1 hour at 10°C wearing only cotton shorts) were measured before and after training. Figure 6-7 shows an improved cutaneous circulation after training as evidenced by the higher skin temperatures. These differences in skin temperature between the pre- and posttraining periods would favor a larger heat loss after training and possibly explain the difference observed in rectal temperature. Finally the differences observed in metabolic rates, shown in Figure 6-8, are possibly related to the changes in body temperatures (3).

Results of this study would then seem to show the beneficial effect of physical fitness on resistance to cold. Indeed the warmer skin temperature would enhance tolerance of the extremities to frostbite and this is made possible through a very slight increase in heat production and similarly quite small drop in deep body temperature. It is interesting to note that the responses of these subjects

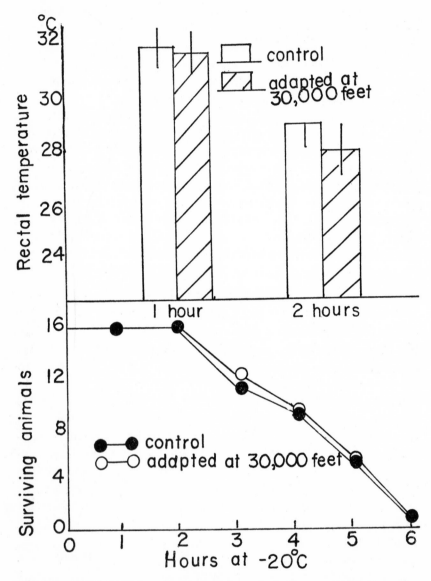

Figure 6-5. Fall in rectal temperature and survival time at −20°C of control rats and rats adapted to altitude by being exposed 18 times at 30,000 feet in a manner described in Figure 6-2. (Unpublished results.)

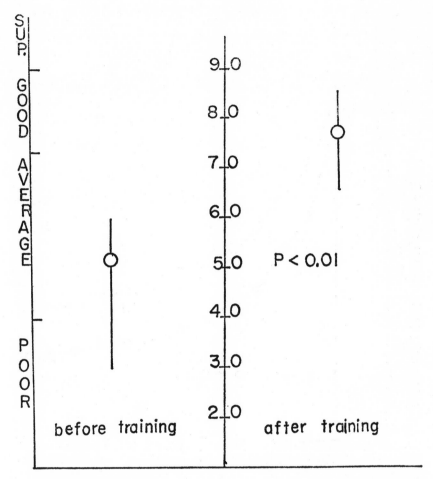

Figure 6-6. Physical fitness scores before and after training program (3).

resemble those reported in Eskimos exposed to cold (Fig. 5-26). Considering the high levels of physical fitness of Eskimos (185, 187), it would seem that physical training might indeed be a very important factor in resistance to cold. Recent studies on laboratory animals by Chin and co-workers (34) have confirmed these results while adding information on the mechanism involved in cross-

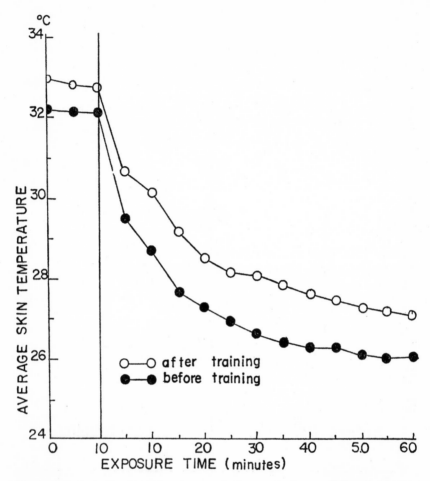

Figure 6-7. Average skin temperatures before and after training in human subjects (3).

adaptation. Groups of rats were trained by different exercise programs. After three weeks the animals were exposed to −20°C for three hours to test the level of tolerance to cold by measurements of colonic temperature. It was found, as shown in Figure 6-9, that the body temperature remained near 37°C in the trained animals at the end of the cold test, while marked hypothermia developed in the control group. A marked increase in plasma adrenaline and nor-

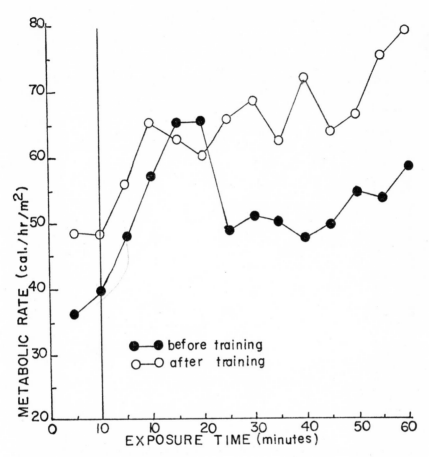

Figure 6-8. Average metabolic rate before and after training in human subjects (3).

adrenaline levels was reported in the same study but this effect was significantly more pronounced in the control than the exercised groups as shown in Figure 6-10 and 6-11. As suggested by the authors, it is quite possible that exercise which increases catecholamine secretion (33), potentiates the sensitivity of the tissues to the calorigenic effect of catecholamines, thus protecting the exercised animals when exposed to cold by reducing the requirements for these amines. Thus it seems that cold exposure (31,

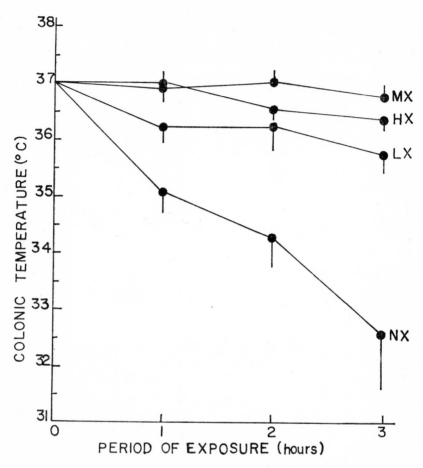

Figure 6-9. Colonic temperatures changes in rats exposed to −20°C for 3 hours. Each bar represents the mean ± standard error for 10 rats. NX = non-exercised control; LX = light exercise; MX = moderate exercise and HX = heavy exercise (34).

152), physical training and chronic injections of catecholamines (138, 146, 187) by increasing plasma levels of catecholamines produce a potentiation of the calorigenic effects of catecholamines which could explain the improved tolerance to cold in these conditions. These findings confirm the importance of the sympathetic nervous system in stress and interstress adaptation and suggest that

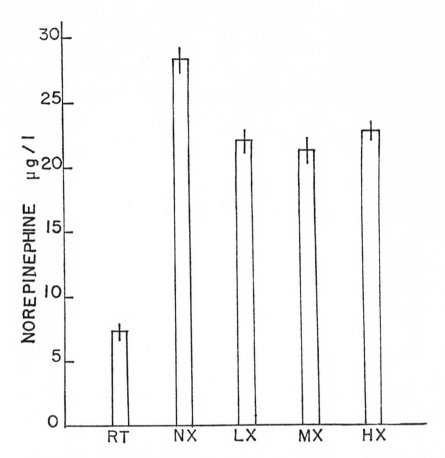

Figure 6-10. Plasma noradrenaline levels in rats exposed to −20°C for 3 hours. See Figure 6-9 for key (34).

this system may be one of the factors common to more than one adaptation as proposed by Adolph (5). Much more work remains to be done in this field and, to quote Folk from his latest book, "it is quite apparent that the future of Environmental Physiology must include many more studies of multifactor environmental influences upon the animal" (63).

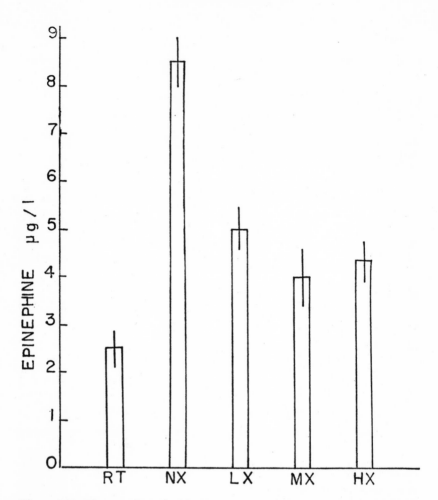

Figure 6-11. Plasma adrenaline levels in rats exposed to −20°C for 3 hours. See Figure 6-9 for key (34).

CHAPTER 7

HYPOTHERMIA, FROSTBITE AND TISSUE PRESERVATION

HYPOTHERMIA

MAMMALS FUNCTION MOST EFFECTIVELY at internal body temperatures varying between 37 and 39°C. Why the body is maintained in this narrow zone is really not known. Of course, this is an ideal temperature at which most enzymatic systems tend to reach a maximum activity. This temperature is maintained through a harmonious control of heat loss and heat production which takes place in the hypothalamic centers. This ideal temperature illustrates the judicious ways through which nature imposes, by some mysterious action, its rigorous laws on living organisms. Indeed 7 degrees lower than 39°C, all mammals lose their control of heat production; below 32°C, hypothermia develops very rapidly and may eventually become irreversible. At 7°C above 39°C brain damage is observed and life becomes impossible with time. The wisdom of nature has set internal body temperature of mammals at around 39°C that is exactly half way between 32 and 46°C which are the extreme temperatures beyond which life itself becomes perilously endangered. Nature, however, is not stagnant. The weakness of the past and the promises of the future are bound in all organisms. By blocking the hypothalamic regulation of body temperature, homeotherms cannot maintain a heat balance even at relatively warm temperature. Figure 7-1 shows that chlorpromazine renders homeotherms similar to poikilotherms. In normal rats exposed to low temperatures extra heat is produced through shivering but when the environmental temperature is too low, heat balance cannot be maintained and body temperature falls. At 32°C

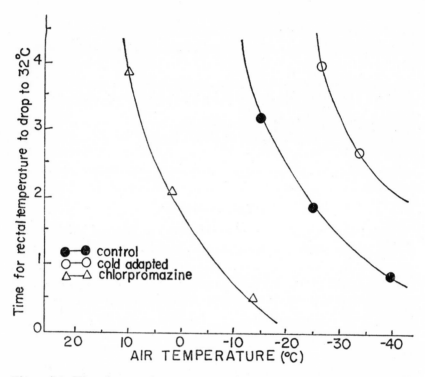

Figure 7-1. Time for rectal temperature of rats to drop to the critical temperature of 32°C at various ambiant temperatures. This process is accelerated by chlorpromazine and delayed by cold adaptation (unpublished results).

this process is accelerated since the animal cannot shiver anymore. With chlorpromazine shivering is not present even at body temperature of 37°C. As a consequence animals given chlorpromazine and exposed to −15°C show an immediate accelerated fall of rectal temperature, the rate of which is comparable to that of normal animal when rectal temperature reaches 32°C. On the other hand this process can be decelerated by adaptation to cold. The animal which has experienced cold exposure is able to use additional ways to resist this stress; the development of nonshivering thermogenesis in cold adaptation is an example of the remarkable capacity that every living organism possesses to expand the limits of its neutrality. Figure 7-2 shows the time it takes for body temperature to reach 32°C when an animal is exposed at different environmental temperatures. For the animal deprived of thermoregulation, by injection of chlorpromazine, after two hours at 0°C body temperature has dropped to 32°C, whereas in cold-adapted animals which have developed nonshivering thermogenesis, this is only observed at —40°C.

[Reduced oxygen requirements of tissues can be produced by hypothermia.]The reduction of body temperature has been successfully obtained by proper use of anesthetics, neuroleptic drugs and increased heat loss by surface cooling. This procedure proved useful and was used in different clinical conditions. Some of the first

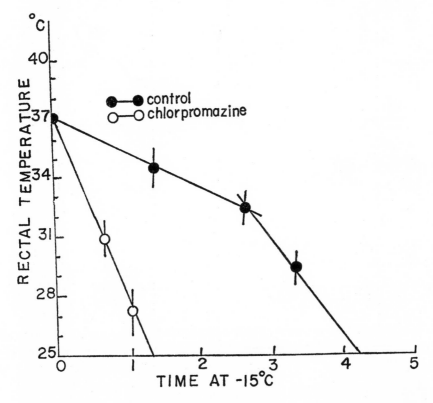

Figure 7-2. Fall in rectal temperature of control and chlorpromazine-treated rats exposed to —15°C for varied periods of time (unpublished results).

studies concerned the use of artificial hibernation for open heart surgery (12, 13, 124). Figure 7-3 shows some cardiovascular changes in a dog cooled to 20°C and rewarmed in a water bath at 40°C, while Figure 7-4 concerns recordings on a forty-three-year-old woman operated for an aneurysm of the internal carotid (17). At this temperature, cardiac output, heart rate and blood pressure are

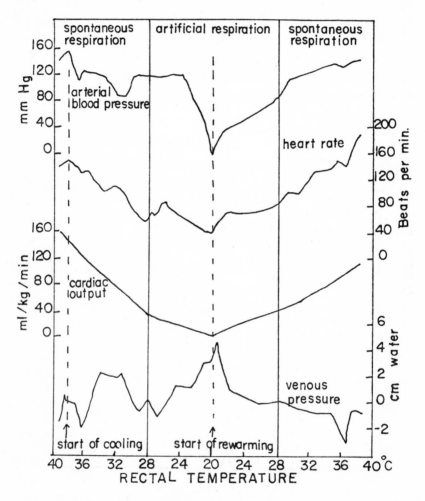

Figure 7-3. Cardiovascular changes in a dog cooled to 20°C with cooling blankets in which an alcohol solution at 1°C was circulated and rewarmed in a water bath at 40°C (13).

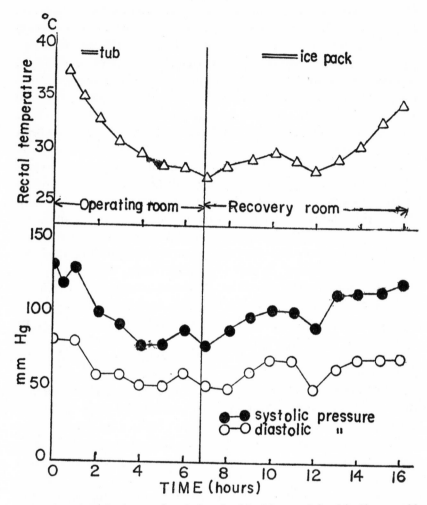

Figure 7-4. Anesthesia was first induced with thiopental in this 43 year old woman. Body temperature was reduced from 37.5 to 32.5°C by immersion in an ice bath for 55 minutes. Body temperature continued to fall to reach approximately 28°C where it remained for about 12 hours. Systolic and diastolic pressures fell in a similar fashion which parallel the drop in rectal temperature (17).

markedly reduced, but this hypothermia is well tolerated by the organism since oxygen requirements in these conditions are also reduced to about 15 percent of their normal requirements (14). At body temperature lower than 20°C, sudden cardiac deterioration may take place. This is characterized by the appearance of ventricular ectopic beats followed by ventricular fibrillation and death. This condition might not be irreversible, however. The generalized peripheral vasoconstriction and the obligatory positive pressure artificial respiration, produce an increased venous pressure which aggravates the congestive failure of a low-output heart. It has been known for some time that venesection improves cardiac output in such conditions (103). Similarly withdrawing of blood was shown effective in reducing venous pressure and increasing cardiac output in hypothermic animals. Venesection in some cases reversed ventricular fibrillation to normal heart action and survival upon rewarming (13). These early studies on laboratory animals have resulted in the use of hypothermia in cardiovascular surgery. Lowered body temperature permits reduced perfusion and even circulatory arrest which greatly facilitate intracardiac surgery. Lowering of body temperature to 15° to 20°C is also used for repair of intracranial aneurysms which are complicated and difficult (162). In 1958 a study was done in which immediate hypothermia was produced in four patients with cardiac arrest exhibiting signs of severe neurological injury (220). Three patients recovered completely and the beneficial effect was related to a reduction of cerebral swelling. This may be considered as a preliminary study which has been reinforced to some extent by results obtained on dogs. Following circulatory arrest of ten minutes, the reduction of body temperature to 32°C for eighteen to thirty-six hours significantly increased the percentage of recovery (220). These studies would indicate that the deleterious effects of cardiac arrest are diminished if the oxygen requirements, especially of the brain, are reduced by lowering internal body temperature. Hypothermia has also been used with some success in septic shock, head trauma, neurosurgery, gastrointestinal bleeding, duodenal ulcer, etc. (170).

FROSTBITE

It is possible to distinguish different phases of freezing injury (25, 35, 175, 189, 198). The first phase is usually accompanied by painful feeling at the level of the cooling tissues. The skin may be pale or more often red due to alternating reflex contraction and dilatation of the vessels known as the "hunting reaction" (154). This is not a damaging phase; it is, however, a critical phase which gives a warning signal. If the exposed part is not protected or re-warmed at this stage, the tissues will be exposed to the second phase of frostbite.

The second phase, depending on the temperature and on the length of exposure, will lead to different degrees of frostbite. During this phase the skin is completely constricted, the pain disappears and the tissues become frozen. If the period of anesthesia is short, that is the time of actual freezing of the skin, superficial layers only will be affected; this is first degree frostbite. The cell damage characterized by leakage of intracellular substances (178) is reversible. The rewarming may be painful, but only for a few minutes. In some parts of the body like the toes, the pain of freezing and rewarming may be absent even in the presence of first degree frostbite. When the tissues are exposed to prolonged freezing at colder temperatures, deeper tissues are damaged and this leads to second, third or fourth degree frostbite which all cause irreversible tissue damage. In the second degree frostbite this damage is repaired by regeneration and connective tissue proliferation, whereas gangrene and permanent tissue loss are observed in fourth degree frostbite.

Many studies have been done on frostbite which has been identified by Blain as the number one problem in military medicine and surgery (15). Other than the actual direct damage to the cell by freezing, the importance of vascular disturbances in the etiology of frostbite has been intensively investigated (40, 126, 173). Essentially before tissue freezing, the constriction is maximal and the blood flow to the tissue is stopped. Upon thawing, considerable hyperemia is present, and because of increased permeability of the vessels, a significant amount of fluid leaks out from the blood. According to Mundth, prior to significant plasma loss, some vascular clumping of platelets, white cells and red cells takes place forming occlusive masses and solid thrombi (175). The length of freezing time is, of course, a dominant factor in determining the degree of injury. Following studies on experimental frostbite, many investigators advocated rapid thawing to minimize tissue injury. Merryman (35) challenged this type of investigation and stated that the beneficial effect of rapid thawing has not been proven because of flaws in experimental design. Another means of diminishing frostbite injury has been to try to maintain circulation following thawing. Different drugs, such as alcohol, heparin, priscolin etc., have been tested to that effect. The difficulty, of course, always consisted in assessing the therapeutic values of any of these substances. Low molecular weight dextran may eventually prove useful in preventing tissue injury; indeed Mundth reported that intravascular aggregation can be prevented and microcirculation improved if animals are given low molecular dextran intraveneously shortly after frosbite (164). Many workers have shown the beneficial effect of dextran in cold injury. More recently Talwar et al. compared the effect of dextran, rapid thawing, low molecular dextran, and sympathectomy as protective means against cold injury (208).

They used groups of ten monkeys. Cold injury was produced on both legs by placing them at $-30°C$ for one hour. One leg was allowed to thaw at room temperature and it served as control. Two or three weeks later the other leg was frozen and immediately submitted to either rapid thawing by immersion in water at $42°C$ for five minutes, to lumbar sympathectomy or to intraveneous injection of 30 ml/Kg of a 10% low molecular dextran (molecular weight 40,000) solution once daily for five days. Figure 7-5 illustrates the outcome of these treatments. Rapid thawing proved useless in preventing tissue damage. Sympathectomy and low molecular dextran very significantly increased the conservation of tissue. These procedures proved useful only when applied immediately upon thawing. These results definitely show that something can be done about frostbite, and to date low molecular dextran is the treatment which has been shown to have the best therapeutic value. Similarly, hyperbaric oxygen was shown to reduce tissue loss from 75 to 25 percent when used immediately after thawing (52).

The best medicine is prevention. One must learn to live in the cold, to dress properly, to eat adequately and to acquire physical fitness.

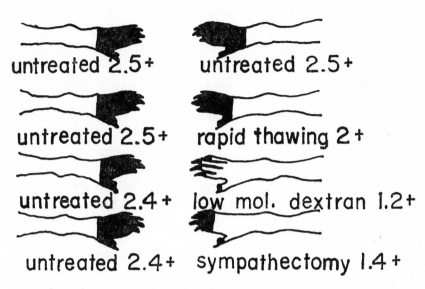

untreated 2.5+   untreated 2.5+

untreated 2.5+   rapid thawing 2+

untreated 2.4+   low mol. dextran 1.2+

untreated 2.4+   sympathectomy 1.4+

Figure 7-5. Average tissue loss on leg of monkeys exposed to −30°C for one hour. Both low molecular dextran and sympathectomy were shown to be significant efficient treatment for the prevention of tissue loss by frostbite (208).

PRESERVATION OF TISSUES BY HYPOTHERMIA

Quite a lot is known about the technology for preservation of food from animal or plant origin. Generally the colder the temperature, the longer the period of conservation. The difficulty arises in the length of time taken in freezing and thawing. The problems become much more important when dealing with human tissues or organs which are frozen for preservation.

When cooling is extremely fast, that is more than −100°C in less than one second, crystals will not form. Liquids become solid by forming an amorphous glass and this in theory would not have deadly effects on cells (116). However, on a practical basis this method is impossible to achieve especially when whole organs are concerned. As a consequence attempts to cool tissues all lead to crystallization. Rapid cooling, however, reduces the size of the crystals and this seemingly reduces the lethality of freezing. Within a certain temperature range, cells are exposed to a thermal shock and become more susceptible to the lethal effect of freezing (200). Lovelock (155) attributes this effect to an increased salt concentration resulting from ice formation within the cell. This expansion of volume within the cell due to expanding ice, increases the tension on the cell membrane which is subjected to disruption. A certain elasticity of cell membrane due to the presence of contracting lipids will oppose to some extent the rupture of the cell (156). The preservation of single cells has been successfully accomplished by using neutral solutes which prevent salt concentration normally produced by freezing. Some studies have shown the successful use of glycerol in preserving viable spermatozoa at different temperatures (155). It was also learned from these studies that the concentration of glycerol has to be sufficiently high to permit access not only to the medium but also within the cell.

Evidence for whole organ preservation by cooling or freezing is much less available. Vacuum dehydration of dog hearts, allowed storage for twenty hours at −8°C and viability up to a maximum of forty hours upon rewarming (8). Another method which has given promising results, consists of perfusing hearts with different substances prior to freezing. As seen in Table 7-I dimethyl sulfoxide and low molecular weight dextran, when perfused in hearts before being frozen at −20°C for twenty minutes, increased survival quite significantly upon thawing (116). These two substances are the more promising solute moderators for tissue preservation. They allow single cell (ova, spermatozoa, red blood cells) freezing for long periods without injury; yet their use in whole organ freezing,

TABLE 7-I
HEARTS FROZEN AT −20°C FOR 20 MINUTES

Solute Moderator	No. Hearts	No. Survivors
Ringers solution	20	0
15% dimethyl sulfoxide	8	6
10% dextran (LMW)	12	5
6% dextran	9	8
12.5% dimethyl sulfoxide in dextran	10	0
15% glycerol-standard series	7	0
15% glycerol "cold" series	8	1

although marked with some success, still remains somewhat limited and can still be considered experimental as far as human organ transplantation is concerned. Similarly more work is needed before the mode of action of these substances is completely understood. As an example of this, both dimethyl sulfoxide, a very hydroscopic liquid, and dextran, a plasma expander, help preserving platelets, yet only the former penetrates inside the cells (188).

ASSESSMENT OF THE ENVIRONMENT BY RESPONSES OF THE FACE

T HE PROBLEM OF EVALUATING the effects of cold on laboratory animals has been discussed in previous chapters. Generally the best method consists of measuring the degree of tolerance of the animal placed in a very cold environment. This is done by recording the fall in body temperature which has been shown to depend on the extent of cold adaptation in the tested animal. Testing the effect of cold and measuring the degree of adaptation in human subjects necessitate other parameters. As was already discussed, the windchill scale has some merit but it is not based on physiological measurements. For instance, when a cold wind is blowing on the face, the windchill value assumes a uniform skin temperature and does not take into consideration the shape of the face or the insulation and vascularization of the different facial area.

EFFECTS OF TEMPERATURE AND WIND ON TEMPERATURE OF THE FACE

Actual measurements of the temperatures of the nose, cheek and forehead at various ambient temperatures and winds serve to illustrate this. The results which are reported in this chapter were obtained on eight male subjects. In all experiments only the face was exposed to the various combinations of wind and air temperature; the rest of the body was always adequately protected and the subject was sitting quietly during the experiments. Figure 8-1 shows the effects of air temperature on skin temperatures of different parts of the face and illustrates the marked response to the wind primarily

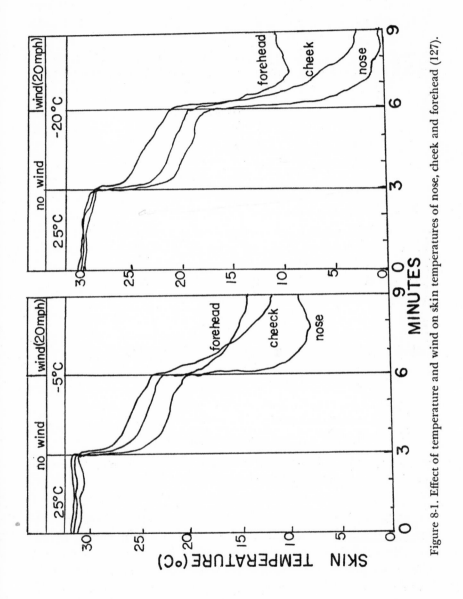

Figure 8-1. Effect of temperature and wind on skin temperatures of nose, cheek and forehead (127).

in the area of the nose. It is well known that the nose is most exposed to cold and subjected to cold injury because of its relative large surface with relation to mass of tissues and because of cold air reaching not only the outside but also the inside nasal surfaces during inspiration. Figures 8-2, 8-3 and 8-4 represent diagrams of variations in skin temperature of the nose, cheek and forehead with relation to various wind speeds and air temperatures. These results

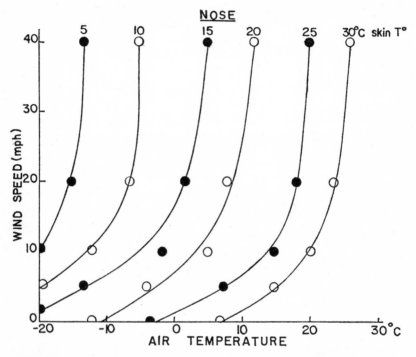

Figure 8-2. Diagram showing the skin temperature of the nose for various combinations of wind and air temperature (127).

were obtained three minutes after exposure to the different environments; as may be seen in Figure 8-1, at this time skin temperature, which shows some variations among individuals, has reached a certain equilibrium. The difference in response between the nose and the forehead is illustrated by the following examples. At −5°C with 20 mph wind or at −12°C with 4 mph wind, the temperature of the nose is 10°C, whereas that of the forehead is 20°C.

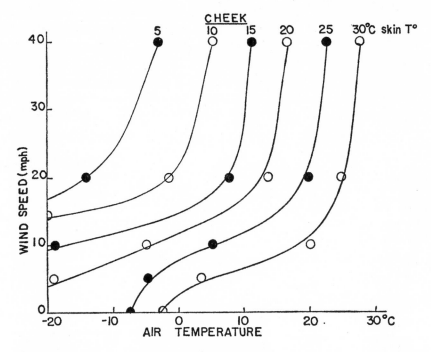

Figure 8-3. Diagram showing the skin temperature of the cheek for various combinations of wind and air temperature (127).

EFFECTS OF TEMPERATURE AND WIND
ON HEART RATE

The effect of cold air on the face with relation of heart rate was discussed in Chapter 4. This is illustrated in Figure 8-5 where subjects were exposed at 20°, 8° or −4°C with and without wind. It can be seen that wind per se at 40 mph causes a drop of approximately 8 beats per minute during expiration; this effect is observed at all temperatures studied. During inspiration, the effect of temperature and wind is still present, but it is greatly reduced as shown in Figure 8-6. The effect of different combinations of temperature and wind on changes in heart rate during expiration are reported in Figure 8-7. A remarkable quantitative response in heart rate is observed with variations of the temperature and wind. As may be expected, these responses parallel those observed for skin tempera-

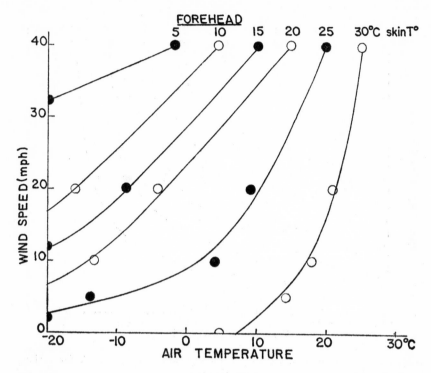

Figure 8-4. Diagram showing the skin temperature of the forehead for various combinations of wind and air temperature.

tures as is shown in Figure 8-8. At this point we have two parameters which permit quantitative evaluation of the cooling effect of the environment on the face: the changes in heart rate and the variations in skin temperatures. These variables are interrelated. The fall in skin temperature stimulates the trigeminal nerve endings and causes the vagal bradycardia reflex. Figure 8-9 shows that this relationship between heart rate and skin temperature is complex. For skin temperatures varying between 35° and 23°C, the correlation between these variables is linear. But for skin temperatures between 23° and 10°C, changes in heart rate are not related to skin temperatures; the heart rate reaches a maximum fall of approximately 10°C at a skin temperature of approximately 23°C.

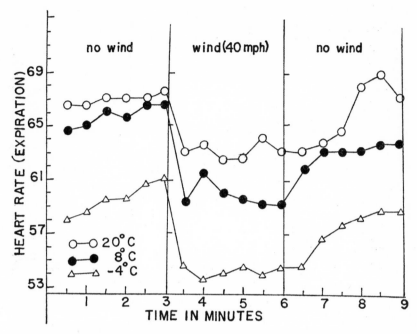

Figure 8-5. Effect of air temperature and wind on heart rate measured during respiratory expiration. Only the face was exposed to the cold wind, the rest of the body being adequately protected (127).

EFFECTS OF TEMPERATURE AND WIND ON THERMAL COMFORT

These results may be explained on the basis of observations made on subjective evaluations of the environmental conditions. In these studies the subjects were asked to evaluate the effect of temperature and wind by chosing a number from zero to one thousand; the subjects were not given any instructions so that their subjective evaluation is strictly personal and relative to their own estimation of comfort. The results obtained are summarized in Figure 8-10. For skin temperature varying between 17° and 32°C, the subjective evaluation of the environment varies between 600 and 850. This is a relative neutral zone which corresponds to sensations varying between comfort and slight discomfort. This is the zone where the heart rate shows the greatest drop (Figure 8-10). In the zone which

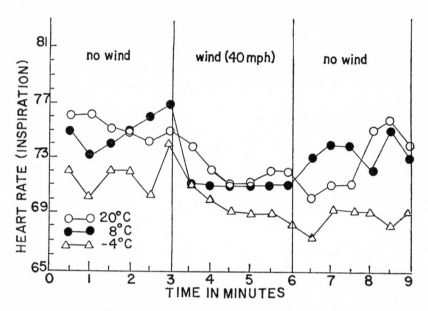

Figure 8-6. Effect of air temperature and wind on heart rate measured during respiratory inspiration. Only the face was exposed to the cold wind, the rest of the body being adequately protected (127).

corresponds to great discomfort, that is for subjective evaluation varying between 600 and 200, variations in skin temperature is very limited and the effect on heart rate stays constant. When the skin temperature is above 32°C, again the subjective evaluation varies between comfort and slight discomfort which is caused this time by the warmer environment. These results are summarized in Figure 8-11. In the zone corresponding to a subjective evaluation varying between 600 and 800, the correlation between the skin temperature, the bradycardia and the subjective evaluation is most evident. However, for subjective evaluation between 600 and 200, the variations in skin temperatures are less pronounced and the bradycardia effect does not become more marked. This zone of discomfort corresponds to an activation of both the sympathetic and parasympathetic nervous systems, as was discussed in Chapter 4. Accordingly the lack of correlation between skin temperatures and heart rate

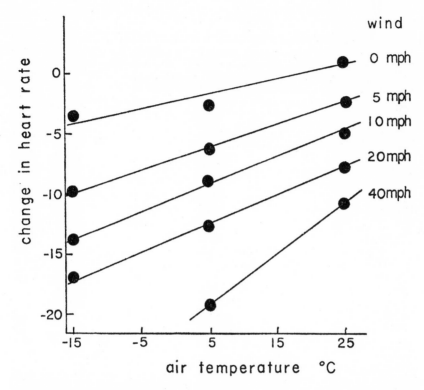

Figure 8-7. Diagram showing the effect of air temperature on heart rate as affected by various wind speeds. Only the face was exposed to the cold wind. the rest of the body being adequately protected (127).

when these temperatures drop below 23°C is possibly due to the fact that the activation of the sympathetic nervous system, which normally causes cardiac acceleration, cancels the vagal brady-cardia action which presumably continues to develop. On the other hand, the increase in heart rate observed when the skin tempera-ture is above 32°C is likely due to some activation of the sympa-thetic nervous system in this zone where some slight discomfort is experienced because of the heat.

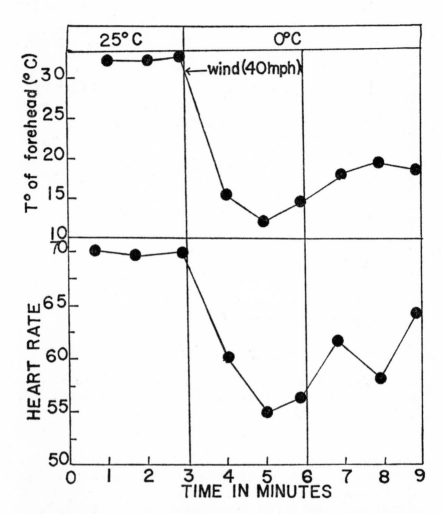

Figure 8-8. Figure showing parallel variations in heart rate and temperature of the forehead with relation to wind and air temperature (127).

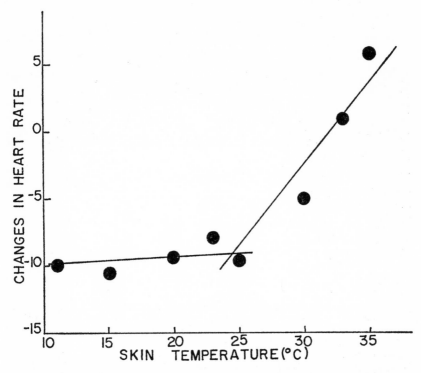

Figure 8-9. Correlation between skin temperature of the forehead and changes in heart rate when the face is exposed to various combinations of temperatures and winds (127).

COMPARISON BETWEEN WINDCHILL INDEX
AND PHYSIOLOGICAL PARAMETERS

By reference to the terminology used in the "windchill" scheme, our results show that for skin temperature below 15°C, where subjective evaluation begins to fall very rapidly, the assessment of the environment may be called "bitterly cold." It is interesting to compare these findings with those reported for the windchill values in Figure 1-6. If the curve of Figure 8-4 corresponding to a skin temperature of the forehead of 15°C is taken to indicate a "bitterly cold" sensation for reasons discussed above, and if this curve is compared to curve number three of Figure 8-12 which also represents

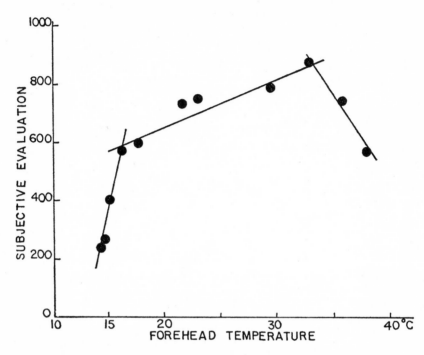

Figure 8-10. Correlation between skin temperature of the forehead and subjective evaluation of the environment when the face is exposed to various combinations of temperatures and winds (127).

the condition so-called "bitterly cold" in the system of windchill value, one can see that at 20 mph wind, for example, the ambient temperature corresponding to "bitterly cold" is −3°C with the first system and −8°C with the second system mentioned. Consequently it would seem that the cooling effect of wind and temperature can be evaluated with some degree of confidence equally well with the windchill system or by actual measurements of skin temperature. The same is true with regard to the nose or cheek except that here the difference in estimates between the two systems is a little more marked than when the forehead is considered. Whatever the system used, a certain variation in the estimates should be expected since both systems rely to some extent on subjective sensations which reflect an overall effect of cold wind on the whole face and not

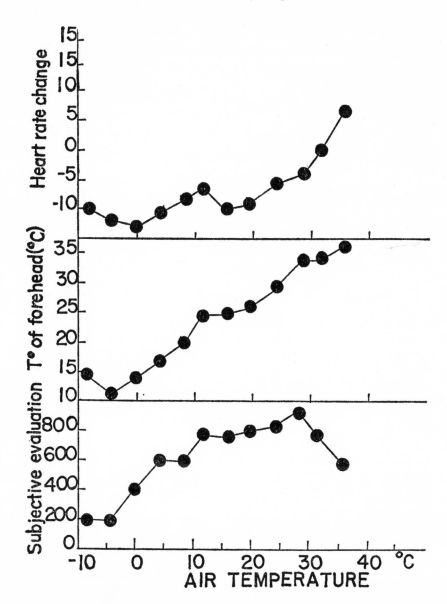

Figure 8-11. Variations in heart rate, temperature of the forehead and subjective evaluation with different air temperature and a constant wind speed at 40 mph. Only the face is exposed to cold, the rest of the body being adequately protected (127).

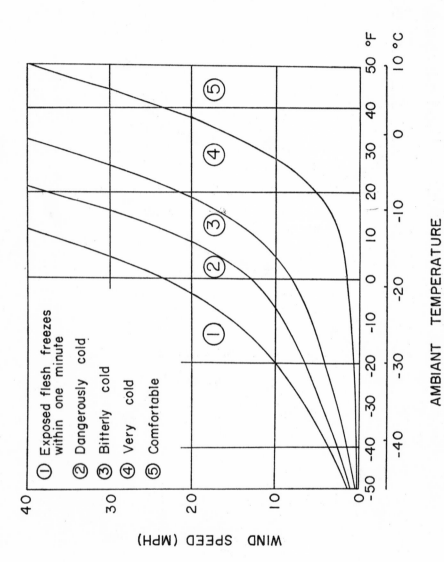

Figure 8-12. Windchill index for various air temperatures and wind speeds (199).

necessarily on one specific area such as the nose or forehead. Just the same, the smaller variations noted with the forehead estimates may be due to the fact that the sensations of cold bitterness seem to be perceived more acutely on the forehead than on the nose or cheek as reported by most subjects.

These parameters offer various possibilities of measuring individual responses to cold and also of assessing the degree of adaptation. The present discussion deals with variations observed in resting conditions. It would be important to study the effect of exercise on the response of the face to cold wind. Possibly these studies would reveal larger variations between the estimates obtained with the windchill system and those calculated from actual measurements of skin temperatures.

REFERENCES

1. Abt, A.F. and Farmer, C.J.: *JAMA, 111*:1555, 1938.
2. Adams, T. and Covino, B.: *J Appl Physiol, 12*:9, 1958.
3. Adams, T. and Heberling, E.J.: *J Appl Physiol, 13*:226, 1958.
4. Adolph, E.F.: In Symposium on Nutrition under Climate Stress., (Eds.): H. Spector and M.S. Peterion. Chicago: QW food and Container Institute, 1954.
5. Adolph, E.F.: In *Handbook of Physiology*. Ed. D.B. Dill, E.A. Adolph and C.G. Wilber. Section 4. Adaptation to the environment. 27-36, 1964.
6. Anderson, T.W., Reid, D.B.W., and Beaton, G.H.: *Can Med Assoc J, 107*:503, 1972.
7. Bacchus, H. and Toompas, C.A.: *Science, 113*:269, 1951.
8. Barsamian, E.M., Jacob, S.W., Collins, S.C., and Owen, O.E.: *Surg Forum, 10*:100, 1959.
9. Beaton, J.R.: *Can J Physiol Pharmacol, 45*:335, 1967.
10. Benzinger, T.H., Kitzinger, C. and Pratt, A.W.: In Hardy, J.D. (Ed.): Temperature: Its measurement and control in science and industry. Reinhold Book Corporation, New York, 1963.
11. Berenson, G.S. and Burch, G.E.: *Am J Med Sci, 233*:45, 1952.
12. Bigelow, W.G., Callaghan, J.C. and Hopps, J.A.: *Ann Surg, 132*:531, 1950.
13. Bigelow, W.G.: *Ann Surg, 132*:849, 1950.
14. Bigelow, W.G., Lindsay, W.K., Harrison, R., Gordon, R.A. and Greenwook, W.F.: *Am J Physiol, 160*:125, 1950.
15. Blain, A. Alex.: *Blain Hosp Bull, 10*:18, 1951.
16. Bligh, J. and Cottle, W.H.: *Experientia, 25*:608, 1969.

17. Boba, A.: *Hypothermia for the Neurosurgical Patient.* Springfield, Thomas, 1960.
18. Boulouard, R.: *Fed Proc, 22*(3), Part 1: 750, 1963.
19. Brown, G.M., Bird, G.S., Boag, T.J., Boag, L.M., Delahaye, J.D., Green, J.E., Hatche, J.D. and Page, J.: *Circulation, 9:*813, 1954.
20. Brück, K. and Wiinnerberg, W.: *Arch Ges Physiol Pfluengers, 290:*167, 1966.
21. Brück, K.: *Brown Adipose Tissue* Elsevier, O. Lindberg (Ed.): New York, 1970.
22. Burch, G.E. and Hyman, A.: *Am Heart J, 53:*665, 1957.
23. Burch, G.E. and De Pasquale, N.P.: *Hot Climates, Man and His Heart.* Springfield, Thomas, 1962.
24. Burn, J.H. and Rand, M.J.: *J Physiol, 147:*135, 1959.
25. Burton, A.C. and Edholm, O.C.: *Man in a Cold Environment.* London, Edward Arnold, 1955.
26. Cadot, M., Julien, M.F. and Chevillard, L.: *Fed Proc, 28:*1228, 1969.
27. Cannon, P. and Keatinge, W.R.: *J Physiol, 154:*329, 1960.
28. Cannon, W.B., Querido, A., Britton, S.W. and Bright, E.M.: *Am J Physiol, 79:*466, 1926.
29. Cannon, W.B.: *The Wisdom of the Body.* New York, Norton, 1932.
30. Cannon, W.B. and Rosenblueth, A.: *Autonomic Neuro-Effector System.* New York, Macmillan, 1937.
31. Carlson, L.D.: *Fed Proc Supp, 5:*25, 1960.
32. Charleston, S.S. and Clegg, K.M.: *Lancet, I:*1401, 1972.
33. Chin, A.K. and Evonuk, E.: *J Appl Physiol, 30:*205, 1971.
34. Chin, A.K., Seaman, R. and Kapilesh-Worker, M.: *J Appl Physiol, 34:* 409, 1973.
35. Whayne, T.F. and DeBakey, M.E. (Eds.): *Cold Injury, Ground Type.* Med. Dept. U.S. Army Govt. Prtg. Office, Washington, 570 p. 1958.
36. Cottle, W.H. and Carlson, L.D.: *Endocrinology, 59:*1, 1956.
37. Cottle, W.H. and Veress, A.T.: *Can J Physiol Pharmacol, 44:*571, 1966.
38. Cottle, W.H., Nash, C.W., Verres, A.T. and Ferguson, B.A.: *Life Sci, 6:*2267, 1967.
39. Cowan, D.W., Diehl, H.S. and Baker, A.B.: *JAMA, 120:*1267, 1942.
40. Crismon, J.M.: *Bull Vas Surg,* 110, 1951.
41. Daniels, Jr., F., Fainer, D.C., Bonmarito, C.L. and Bass, D.E.: *Fed Proc, 10:*32, 1951.
42. Davis, T.R.A. and Johnston, S.R.: *J Appl Physiol, 16:*221, 1961.
43. Dawkins, M.J.R. and Hull, D.: *J Physiol, (Lond), 172:*216, 1964.
44. De Pasquale, N.P. and Burch, G.E.: *Am J Med Sci, 242:*468, 1965.
45. Desmarais, A. and LeBlanc, J.: *Can J Med Sci, 30:*157, 1952.
46. Depocas, F., Hart, J.S. and Héroux, O.: *J Appl Physiol, 10:*393, 1957.
47. Depocas, F. and Masironi, R.: *Am J Physiol, 199:*1051, 1960.
48. Depocas, F.: *Can J Biochem Physiol, 38:*107, 1960.

49. Depocas, F.: *Fed Proc Symposium on Temperature, 19:*23, 1960.
50. Dosne de Pasqualini: *Am J Physiol, 147:*598, 1946.
51. Dosne, C.: *Rev Soc Argentina Biol, 23:*234, 1947.
52. Dritsas, K.G. and Kowalewski, K.: *Can J Surg, 11:*85, 1968.
53. Dugal, L.P. and Thérien, M.: *Can J Res Sect. E, 25:*111, 1947.
54. Dugal, L.P.: Cold injury. Trans. 2nd Conf. 1952, Josiah Macy Jr Foundation.
55. Dupont, A., Bastarache, E., Endröezi, E. and Fortier, C.: *Can J Physiol Pharmacol, 50:*364, 1972.
56. Eagan, C.J.: *Fed Proc Symposium on Temperature Acclimation, 22:*947, 1963.
57. Elmadjian, F., Hope, J.M. and Lamson, E.T.: *Recent Progr Horm Res, 14:*513, 1958.
58. von Euler, U.S.: *Noradrenaline.* Springfield, Thomas, 1954.
59. von Euler, U.S. and Lishanko, F.: *Acta Physiol Schand, 45:*122, 1959.
60. von Euler, U.S.: *Neuroendocrinology,* Vol. II, Martini, L. and Ganong, W.F. (Eds.) Academic Press, New York 1967.
61. Feldberg, W. and Myers, R.D.: *J Physiol, 173:*226, 1964.
62. Feldberg, W.S.: In Hardy, Gagge, and Stolwijk (Eds.): *Physiological and Behavioral Temperature Regulation.* Springfield, Thomas, 1970.
63. Folk, G.E.: *Introduction to Environmental Physiology.* Philadelphia, Lea and Febiger, 1974.
64. Fortier, C., Harris, G.W., McDonald, I.R.: *J Physiol (Lond), 136:*344, 1957.
65. Fortier, C.: *Second Int Congress of Psycho-neuroendocrinology.* Hungarian Academy of Sciences, 1972.
66. Frankenhaeuser, M., Sterky, K. and Jaerpe, G.: *Percept Motor Skills, 15:*63, 1962.
67. Franz, W.L., Sands, G.W. and Heyl, H.L.: *JAMA, 162:*1224, 1956.
68. Fregley, M.J.: *Am J Physiol, 176:*267, 1954.
69. Galton, V.A. and Nisula, B.C.: *Endocrinology, 85:*79, 1969.
70. Gellhorn, E. and Loofbourrow, G.N.: *Emotions and Emotional Disorders.* New York, Harper, 1963.
71. Gilgen, A., Maickel, R.P., Nikodijevic, O. and Brodie, B.B.: *Life Sci, 1:*709, 1962.
72. Glasser, E.M.: *Int J Bioclimat, 1:*1, 1957.
73. Glasser, E.M. and Whittow, G.C.: *J Physiol (Lond), 136:*98, 1957.
74. Glasser, E.M. and Griffin, J.P.: *J Physiol (Lond), 160:*429, 1962.
75. Glazebrook, A.J. and Thomson, S.: *J Hyg, 42:*1, 1942.
76. Goldby, F.C., Hicks, C.S.: O'Connor, W.J. and Sinclair, D.A.: *Aust J Exp Biol Med Sci, 16:*29, 1938.
77. Grant, R.T. and Bland, E.F.: *Heart, 15:*385, 1931.
78. Griffin, J.: *Clin Sci, 24:*129, 1963.
79. Hammel, H.T., Elsner, R.W., Le Messurier, D.H., Andersen, H.T. and

Milan, F.A.: *J Appl Physiol, 14:*605, 1959.

80. Hammel, H.T., Elsner, R.W., Andersen, K.L., Scholander, P.F., Coon, C.S., Medina, A., Strozzie, L., Milan, F.A. and Hock, R.J.: Tech. Report no. 60-633, Wright Air. Development Division, 1960.

81. Hammel, H.T., Hildes, J.A., Jackson, D.C. and Andersen, H.T.: Tech. report no. 62-44, Arctic Aeromedical Laboratory, Ladd AFB, 1962.

82. Hammel, H.T.: *Physiological Controls and Regulations.* In Yamamoto and J. R. Brobeck (Eds.), Philadelphia, Saunders, 1965.

83. Hannon, J.P., Evonuk, E. and Larson, A.M.: *Fed Proc, 22:* Part I 783, 1963.

84. Hannon, J.P. and Larson, A.M.: *Am J Physiol, 203:*1055, 1962.

85. Harding, R.G., Roman, D. and Whelan, R.F.: *J Physiol, 181:*401, 1965.

86. Harley, A., Starmer, C.F. and Greenfield, J.C.: *J Clin Invest, 48:*895, 1919.

87. Harris, W.S., Schoenfeld, C.D., Brooks, R.H.: *Am J Cardiol, 17:*484, 1966.

88. Hart, J.S. and Jansky, L.: *Can J Physiol Pharmacol, 41:*629, 1963.

89. Heim, T., and Hull, D.: *J Physiol (Lond), 187:*271, 1966.

90. Hellon, R.: *Essays on Temperature Regulation.* Bligh and Moore (Eds.): North Holland, Amsterdam, 1972.

91. Hemingway, A., Forgrave, P. and Birzis, L.: *J Neurophysiol, 17:*375, 1954.

92. Hensel, H., Iggo, A .and Witt, I.: *J Physiol, 153:*113, 1960.

93. Hensel, H.: In *Temperature its Measurement and Control in Science and Industry.* J.D. Hardy (Ed.), Reinhold Book Corporation, New York, 1963.

94. Héroux, O. and Petrovic, V.M.: *Can J Physiol Pharmacol, 47:*963, 1969.

95. Héroux, O.: *Fed Proc, 28:*955, 1969.

96. Herrington, L.P.: *Am J Physiol, 129:*123, 1940.

97. Hick, C.S., Matters, R.F. and Mitchell, M.L.: *Aust J Exp Biol Med Sci, 8:*69, 1931.

98. Hildes, J.A.: *Fed Proc Symposium on Temperature Acclimation, 22:*843, 1963.

99. Himms-Hagen, J.: *J Physiol (Lond), 205:*393, 1969.

100. Himms-Hagen, J.: *Lipids, 7:*310, 1972.

101. Hines, E.A. and Brown, G.E.: *Proc Staff Meeting Mayo Clinic, 7:*322, 1932.

102. Hong, S.K.: *Fed. Proc. Symposium on Temperature Acclimation, 22:* 831, 1963.

103. Howarth, S., McMichal, J., and Sharpey-Schafer, E.P.: *Clin Sci, 6:*41, 1946.

104. Horvath, S.M., Freedman, A. and Golden, H.: *Am J Physiol, 150:*99, 1947.

105. Horwitz, B.A. and Detrick, J.F.: *Fed Proc, 30:*319, 1971.

106. Hsieh, A.C.L. and Carlson, L.D.: *Am J Physiol, 190:*243, 1957.
107. Irving, L., Scholander, P.F. and Grincell, S.W.: *Am J Physiol, 135:*557, 1941.
108. Itoh, S.: *Physiology of Cold-Adapted Man.* Sapporo, Hokkaido University School of Medicine, 1974.
109. Jansky, L.: *Fed Proc Int Symposium on Temperature, 25:*1297, 1966.
110. Jansky, L.: Bartunkova.D3e
110. Jansky, L., Bartunkova, R. and Zeisberger, E.: *Physiol Bohemoslovaca, 16:*366, 1967.
111. Jansky, L.: *Fed Proc Symposium on Temperature, 28:*1054, 1969.
112. Jenec, V.: *Can J Physiol Pharmacol, 42:*585, 1964.
113. Jobin, M., Koch, B. and Fortier, C.: *Fed Proc, 25:*516, March-April 1966.
114. Johnson, R.E. and Kark, R.M.: *Science, 105:*378, 1947.
115. Karow, A.M. and Webb, W.R.: *Surg Gynecol Obstet, 119:*609, 1964.
116. Karow, A.M. and Webb, W.R.: *Arch Surg, 91:*572, 1965.
117. Kawhata, A. and Adams, T.: *Proc Soc Exp Biol Med, 106:*862, 1961.
118. Keatinge, W.R.: *J Physiol, 153:*166, 1960.
119. Keatinge, W.R.: *Survival in Cold Water.* Oxford, Blackwell Scientific Publications, 1969.
120. Keys, A.: *Nut Abst Rev, 19:*1, 1949-50.
121. Kirk, J.E. and Chieffi, M.: *Gerontology, 8:*301, 1953.
122. Koch, B., Jobin, M., Dulac, S. and Fortier, C.: *Can J Physiol Biochem, 50:*360, 1972.
123. Kreyberg, L.: *Physiol Rev, 29:*156, 1949.
124. Laborit, H. and Huguenard, P.: *J Chir (Paris), 67:*631, 1951.
125. Labrie, F.: Thesis, Québec, Université Laval, 1967.
126. Lake, N.: *Lancet,* I.I., Oct. 13th, 1917.
127. LeBlanc, J., Barabé, J., Blais, B., Côté, J. and Dulac, S.: Unpublished results.
128. LeBlanc, J., Côté, J., Savary, P., Girard, B., Beaudoin, R. and Drolet, N.: Unpublished data.
129. LeBlanc, J.: Unpublished data.
130. LeBlanc, J.: *Can J Biochem Physiol, 32:*354, 1954.
131. LeBlanc, J., Stewart, M., Marier, G. and Whillians, M.G.: *Can J Biochem Physiol, 32:*407, 1954.
132. LeBlanc, J.: DRNL report no. 3/55, 1954. Defence Research Board of Canada, Ottawa.
133. LeBlanc, J.: *J Appl Physiol, 9:*395, 1956.
134. LeBlanc, J.: *J Appl Physiol, 10:*281, 1957.
135. LeBlanc, J., Hildes, J.A., Héroux, O.: *J Appl Physiol, 15:*1031, 1960.
136. LeBlanc, J.: *J Appl Physiol, 17:*950, 1962.
137. LeBlanc, J., Pouliot, M. and Rhéaume, S.: *J Appl Physiol, 19:*9, 1964.
138. LeBlanc, J. and Pouliot, M.: *Am J Physiol, 207:*853, 1964.

139. LeBlanc, J. and Pouliot, M.: *Am J Physiol, 207:*854, 1964.
140. LeBlanc, J.: *Ann NY Acad Sci, 104:*721, 1966.
141. LeBlanc, J. and Potvin, P.: *Can J Physiol Pharmacol, 44:*289, 1966.
142. LeBlanc, J.: *Am J Physiol, 212:*530, 1967.
143. LeBlanc, J., Robinson, D., Sharman, D.F. and Tousignant, P.: *Am J Physiol, 213:*1419, 1967.
144. LeBlanc, J., Villemaire, A. and Vallières, J.: *Arch Int Physiol Biochem,* 731, 1969.
145. LeBlanc, J.: Federation Proceeding. *Symposium on Cold Adaptation, 28:*996, May-June, 1969.
146. LeBlanc, J. and Villemaire, A.: *Am J Physiol, 218:*1742, 1970.
147. LeBlanc, J., Roberge, C., Vallières, J. and Oakson, G.: *Can J Physiol Pharmacol, 49:*96, 1971.
148. LeBlanc, J.: *Nonshivering Thermogenesis.* Symposium held in Prague. Academia, 1971.
149. LeBlanc, J., Vallières, J. and Vachon, C.: *Am J Physiol, 222:*1044, 1972.
150. LeBlanc, J.: In *Polar Human Biology.* Edholm, O.G. (Ed.), London, Heinemann, 1973.
151. Leblond, C.P., Gross, J., Peacock, W. and Evans, R.D.: *Am J Physiol, 140:*671, 1944.
152. Leduc, J.: *Acta Physiol Scand, 53,* Suppl. 183, 1, 1961.
153. Leduc, J. and Rivest, P.: *Rev Can Physiol, 28:*49, 1969.
154. Lewis, T.: *Heart, 15:*177, 1930.
155. Lowelock, J.E. and Polge, C.: *Biochem J (Lond), 58:*618, 1954.
156. Lowelock, J.E.: *Br J Haematol, 1:*117, 1955.
157. Maikel, R.P., Westermann, E.O. and Brodie, B.B.: *J Pharmacol Exp Ther, 134:*167, 1961.
158. Maickel, R.P., Sussman, H., Yamada, K. and Brodie, B.: *Life Sci, 2:* 210, 1963.
159. Mallow, S.: *Am J Physiol, 204:*157, 1963.
160. Masironi, R. and Depocas, F.: *Can J Biochem Physiol, 39:*219, 1961.
161. Metzger, C.C., Chough, C.B., Kroetz, F.W. and Leonard, J.J.: True isovolumic contraction time. *Am J Cardiol, 25:*434, 1970.
162. Michenfelder, J.D., Terry, H.R., Daw, E.F. and Vihlein, A.: *Surg Clin North Am, 45:*(4) 889, 1965.
163. Moore, R.E. and Simmonds, M.A.: *Fed Proc Symposium on Temperature, 25:*1329, 1966.
164. Mundth, E.D., Long, D.M. and Brown, R.B.: *Trauma, 4:*246, 1964.
165. Myers, R.D. and Sharpe, L.C.: *Science, 161:*572, 1968.
166. Myers, R.D.: *The Hypothalamus.* Haymaker, Anderson, and Nauta. (Eds.) Springfield, Thomas, 1969.
167. Myers, R.D.: *Physiological and Behavioral Regulation.* Hardy, Gagge and Stolwijk (Eds.) Springfield, Thomas, 1970.
168. Meyers, R.D. and Beleslin, D.B.: *Am J Physiol, 220:*1846, 1971.

169. Nair, C.S., Malhotra, S., Tiwari, O.P. and Gopinath, P.M.: *Aerospace Med,* 42:991, 1971.
170. Nealon, T.F. and Gosin, S.: *Med Clin North Am,* 49:(5) 1181, 1965.
171. Olmsted, J.M.D. and Olmsted, E.H.: *Claude Bernard and the Experimental Method in Medicine.* New York, Abelard, 1952.
172. Pagé, E. and Babineau, L.M.: *Rev Can Biol,* 9:202, 1950.
173. Pauling, L.: *Vitamin C and the Common Cold.* San Francisco, W.H. Freeman, 1970.
174. Pavlov, I.P.: *Lectures on Conditioned Reflexes.* Trans. by W. Horsley. New York, Gantt, 1928.
175. *Proceedings of Symposium on Frotbite, 1964.* Arctic Aeromedical Laboratory, Fort Wainwright, Alaska.
176. Raab, W., Paula e Silva, P. de and Starcheska, Y.K.: *Cardiologia, 33:* 350, 1958.
177. Radomski, M.W. and Orme, T.: *Am J Physiol,* 220:1852, 1971.
178. Régnier, E.: *Rev Allergy,* 22:835, 1968.
179. Rennie, D.W., Covino, B.G., Howell, B.J., Song, S.H., Kang, B.S. and and Hong, S.K.: *J Appl Physiol,* 17:961, 1962.
180. Rennie, D.W.: *Fed Proc Symposium on Temperature Acclimation, 22:* 828, 1963.
181. Richet, C.: *Arch Biol,* 5:312, 1893.
182. Ritzel, G.: *Helv Medica Acta,* 28:63, 1961.
183. Rodahl, K.: *North: The Nature and Drama of the Polar World.* New York, Harper, 1953.
184. Rodahl, K.: *Tech. Rep.* 57-21 Arctic Aeromedical Laboratory, Ladd Air Force Base, Alaska, 1957.
185. Rodahl, K. and Issekutz, B.: Nutritional requirements in the cold. Symposia on Arctic Biology and Medicine. Arctic Aeromedical Laboratory, Fort Wainwright, Alaska, 1965.
186. Rode, A. and Shephard, R.: *J Med Sci Sports,* 5:170, 1973.
187. Rowell, L., Brengelmann, G.L. and Murray, J.A.: *J Appl Physiol, 27:* 673, 1965.
188. Roy, A.J. and Djersassi, I.: *Cryobiology,* (Suppl 1), *1:*20, 1964.
189. Sapin-Jaloustre: *Enquête sur les Gelures.* Paris, Hermann et Cie éditeurs, 1956.
190. Scaria, K.S., Prabha, K.: *Ind J Exp Biol,* 5:255, 1967.
191. Sellers, E.A., You, S.S. and Thomas, N.: *Am J Physiol,* 165:481, 1951.
192. Selye, H.: *Stress.* Montréal, Acta Inc., 1950.
193. Selye, H.: *The Physiology and Pathology of Exposure to Stress.* Montréal, Acta Inc., 1950.
194. Scholander, P.F., Hock, R., Walters, V., Johnson, F. and Irving, L.: *Biol Bull,* 99:225, 1950.
195. Scholander, P.F., Hammel, H.T., Hart, J.S., LeMessurier, D.H. and Steen, J.: *J Appl Physiol,* 13:211, 1958.

196. Schönbaum, E., Johnson, G.E., Sellers, E.A. and Gill, M.J.: *Nature, 210:* 426, 1966.
197. Shore, P.Q.: *Pharmacol Rev, 14:*531, 1962.
198. Schuman, L.M.: Cold Injury 2nd Conference, 1952, The Josiah Macy Jr. Foundation, New York.
199. Siple, P.A. and Passel, C.F.: *Proc Am Philos Soc, 89:*177, 1945.
200. Smith, A.V.: *Biological Applications of Freezing and Drying.* Harris, R.J.C. (Ed.) New York. Academic Press, 1954.
201. Smith, R.E. and Moijer, D.J.: *Physiol Rev, 42:*60, 1962.
202. Smith, R.E. and Roberts, J.C.: *Am J Physiol, 206:*143, 1964.
203. Smith, R.E. and Horwitz, B.A.: *Physiol Rev, 49:*330, 1969.
204. Starr, P. and Rosbelley, R.: *Am J Physiol, 130:*549, 1940.
205. Stefanssen, V.: *The Friendly Arctic.* New York, Macmillan, 1953.
206. Stein, H.J., Bader, R.A., Eliot, J.W. and Bass, D.E.: *J Clin Endocrinol, 9:*529, 1949.
207. Stewart, C.P., Learmouth, J.R. and Pollock, G.A.: *Lancet, 240:*818, 1941.
208. Talwar, J.R., Gulati, S.M. and Kapur, B.M.L.: *Ind J Med Res, 59:*242, 1917.
209. Theriault, D.G., Hubbard, R.W. and Mellin, D.B.: *Lipids, 4:*413, 1969.
210. Theriault, D.G., Morningstar, J.F. and Winters, S.V.G.: *Life Sci, 8:* 1353, 1969.
211. Theriault, D.G. and Mellin, D.B.: *Lipids, 6:*486, 1971.
212. Thérien, M., LeBlanc, J., Héroux, O. and Dugal, L.P.: *Can J Res (Sect E), 27:*349, 1949.
213. Trendelenburg, U.: *Pharmacol Rev, 15:*225, 1963.
214. Ungar, G.: *Lancet, 244:*421, 1943.
215. Vallières, J., Bureau, M. and LeBlanc, J.: *Can J Physiol Pharmacol, 50:*576, 1972.
216. Villemaire, A.: Études sur l'interaction de la thyroxine et des catecholamines. Ph. D. Thesis, Laval University, Québec, 1970.
217. Vogt, M.: *J Physiol, 123:*451, 1954.
218. Weissler, A.M., Peeler, R.G. and Roehll, W.H.: *Am Heart J, 62:*367, 1961.
219. Welch, B.E., Buskirk, E.R., Mann, J.B., Innell, W., Friedmann, T.E., Kreider, M., Brebbia, R., Morana, N. and Daniels, F.: Report no. 173, Army Medical Nutrition Laboratory, Denver, 1955.
220. Williams, G.R. and Spencer, F.C.: *Ann Surg, 148:*462, 1958.
221. Wilson, C.W.M. and Low, H.S.: *Acta Allergololica, 24:*367, 1970.
222. Wolf, S. and Hardy, J.D.: *J Clin Invest, 20:*521, 1941.
223. Woods, R. and Carlson, L.D.: *Endocrinology, 59:*323, 1956.
224. Wyndham, C.H. and Morrison, J.F.: *J Appl Physiol, 13:*219, 1958.
225. Zotterman, Y.: *Handbook of Physiology: Neurophysiology.* Am. Physiol. Soc, Washington D.C., 1959.

SUBJECT INDEX

A

Aborigines of Australia, 116, 118
Adaptation, 90-145
  and autonomic nervous system, 91
  and cold tolerance, 148, 149, 154
  and noradrenaline, 54, 58
  and survival, 112
  by prolonged exposures, 91-104
  by short exposures, 104-116
  in human, 116-140
  to altitude, 147, 150
Adaptive responses, vii
Adenyl cyclase, 32, 97
Adrenal cortex, 58, 60
Adrenal medulla, 113
Adrenaline
  effect on metabolism, 58, 59, 96
  excretion, 91, 93
  perfusion, 73
  plasma levels in cold, 156
  secretion, 160
Ainu, 120
Alacalup Indians, 120
Alcohol and frostbite, 168
Alpha receptors, 99
Ama, 120
Angor pectoris, 76
Arctic, 128
Arctic mammals, 3
Autonomic nervous system, 74, 80

B

Basal Metabolic Rate, 25, 54
Behavioral Adaptation, 146
Beta receptor, 97, 99
Blood pressure, 65, 69, 82, 134, 140
Bradycardia
  and face stimulation, 69-84
Breath holding, 81
Brown adipose tissue, 32-35
Bushmen of Kalahari Desert, 117

C

Calorigenesis, 100
Cardiac acceleration, 179
Cardiac output, 166
Cardiovascular effects
  noradrenaline, 100
Catecholamines, 27
  see noradrenaline
Cheek temperature, 172-174
Chlorpromazine, 162
Clo unit, 4
Clothing, 9
Cold
  and altitude, 147
  and exercise, 153
  immersion, 20, 133
  induced vasodilatation, 65
  pain, 65
  pressor test, 65, 76
  resistance, 93
  tolerance, 20, 22, 102, 133, 153
  water test, 140
Colonic temperature, 98
Common cold, 40
Conditioning to cold, 137
Continuous exposure
  to moderate cold, 116, 150
Cutaneous circulation, 153
Cyclic AMP, 97

D

Dextran
  and frostbite, 168-170
Dimethyl sulfoxide
  and frostbite, 170

E

Electro-mechanical time
  QS_2, 69
Emotions
  and adaptation, 113

Epididymal fat
  and cold resistance, 27
Eskimos
  adaptation 8, 116, 126-130
  and cold pressor test, 132, 137
  hand blood flow, 64
  physical fitness, 155
Extremities
  and cold exposure, 128

F

Face immersion
  and bradycardia, 80
  and cold assessment, 172-185
  systolic time intervals, 69-73
Fishermen
  of Gaspé, 134-138
Food intake, 38, 39
Foot trouble
  and vitamin C, 45, 48
Forehead
  temperature, 172-174
Free fatty acids, 173-174
Frostbite, 129, 161, 166, 167

G

Gaspé fishermen, 133
Glucocorticoid secretion, 59
Gluconeogenesis, vii
Glucose metabolism, 27
Guanethidine, 56

H

Habituation
  in humans, 137, 138, 146-147
  in rats, 107, 113
Hand blood flow, 64, 129
Hand immersion test, 73, 74
Heart attacks, 82
Heart rate, 68, 175, 178, 180, 183
Heparin
  and frostbite, 168
Hippocampus, 62
Hot climate, 88
Humid environment, 87
Hunting reaction, 65, 167
Hypophyso-adrenal system, 27
Hypothalamus, 11, 14, vii

Hypothermia
  and drugs, 161-169
  and exercise, 156

I

Insulation, 3
Intracardiac surgery, 166
Intermittent exposure
  to severe cold, 104, 116, 126
Interstress adaptation, 146
Isoproterenol, 97, 98, 100, 101

K

Kalahari Bushmen, 119, 126
Ketone bodies, 120

L

Left ventricular ejection time, 69
  LVET
Lethal temperature, 3
Lipase activity, 32, 98
Lipid metabolism, 27
Lipolysis, 32

M

Man
  and cold, vii
Metabolism
  and noradrenaline, 92
Metabolic rate
  and exercise, 157

N

Neutral zone, 177
Non Shivering thermogenesis
  and adaptation, 25, 29
  and noradrenaline, 52, 91
  in humans, 125, 126
Non-Specific responses, 150
Noradrenaline
  and adaptation, 91-110
  brown adipose tissue, 32, 35
  in exercise, 156, 159
  in hypothalamus, 15
  injections, 93
  response in humans, 120, 125
  respone in rats, 52-59
  secretion, 58

sensitivity, 147
systolic time intervals, 73

P

Pain sensation, 144
Parasympathetic system, 79
Physical activity, 22, 23
Physical fitness
and resistance to cold, 153, 155
Pituitary gland, 60
Pre-ejection period
PEP, 69
Preservation of tissues
by cold, 169
Primitive populations, 123

Q

QRS Complex, 69

R

Rapid cooling, 169
Receptors, 11
Rectal temperature, 136
Repeated exposure
to severe cold, 150
Repeated immersion, 104
Reserpine, 27
Reistance to cold, 90, 153
Resistance to heat, 87
Respiratory expiration, 177
Respiratory inspiration, 178
Respiratory quotient, 27
Response to cold, vii

S

Sensitivity
to noradrenaline, 91, 99, 107
Serotonin, 15
Shivering
and adaptation, 162
central control, 10, 18
in Eskimos, 137
in fishermen, 136, 137
in humans, 123, 125
Short exposure
to severe cold, 104
Size of the body, 15
Skin heat flow, 174, 176

Skin receptors, 12, 72
Skin temperature, 133-136, 156
Subcutaneous fat, 17, 20, 127, 128
Subjective evaluation, 182, 183
Sudden death
and cold, 81
Survival time, 109, 111, 112, 149, 154
Sympathetic nervous system, 79, 91, 100
Sympathectomy
and cold tolerance, 168
Systolic time intervals, 69, 72, 73, 131,
132, 142

T

Temperature
of finger, 130
of forehead, 180, 182
of nose, 172
Thermal comfort, 177
Thyroxine, 50-56, 100, 103, 113, 115
Thyroxine secretion, 50, 52
Tissue injury, 168
Tissue insulation, 127
Tissue preservation, 161
Tolerance
to altitude, 152
to cold, 114, 150
Trigeminal nerve, 70
Triiodothyroxine, 52
TSH, 51, 60

U

Uptake of noradrenaline
in cold adaptation, 100

V

Vagal bradycardia, 70, 72, 176
Ventricular contraction, 101
Ventricular fibrillation, 166
Vitamin C, 40-46

W

Wind effects, 172, 175, 177
Wind speed, 179
Wind velocity, 8
Windchill index, 9, 181, 184, 185

Z

Zone of discomfort, 178